Maternal Control of Development in Vertebrates

My Mother Made Me Do It!

Developmental Biology

Editors

Daniel S. Kessler, *University of Pennsylvania School of Medicine*

Developmental biology is in a period of extraordinary discovery and research in this field will have a broad impact on the biomedical sciences in the coming decades. Developmental Biology is interdisciplinary and involves the application of techniques and concepts from genetics, molecular biology, biochemistry, cell biology, and embryology to attack and understand complex developmental mechanisms in plants and animals, from fertilization to aging. Many of the same genes that regulate developmental processes underlie human regulatory gene disorders such as cancer and serve as the genetic basis of common human birth defects. An understanding of fundamental mechanisms of development is providing a basis for the design of gene and cellular therapies for the treatment of many human diseases. Of particular interest is the identification and study of stem cell populations, both natural and induced, which is opening new avenues of research in development, disease, and regenerative medicine. This eBook series is dedicated to providing mechanistic and conceptual insight into the broad field of Developmental Biology. Each issue is intended to be of value to students, scientists, and clinicians in the biomedical sciences.

Maternal Control of Development in Vertebrates: My Mother Made Me Do It!
Florence L. Marlow
www.morganclaypool.com

ISBN: 9781615040513 paperback

ISBN: 9781615040520 ebook

DOI: 10.4199/C00023ED1V01Y201012DEB005

A Publication in the Morgan & Claypool Publishers Life Sciences series

DEVELOPMENTAL BIOLOGY

Book #5

Series Editor: Daniel Kessler, Ph. D., University of Pennsylvania

Series ISSN

ISSN 2155-3521 print

ISSN 2155-353X electronic

Maternal Control of Development in Vertebrates

My Mother Made Me Do It!

Florence L. Marlow
Albert Einstein College of Medicine

DEVELOPMENTAL BIOLOGY #5

MORGAN & CLAYPOOL LIFE SCIENCES

ABSTRACT

Eggs of all animals contain mRNAs and proteins that are supplied to or deposited in the egg as it develops during oogenesis. These maternal gene products regulate all aspects of oocyte development, and an embryo fully relies on these maternal gene products for all aspects of its early development, including fertilization, transitions between meiotic and mitotic cell cycles, and activation of its own genome. Given the diverse processes required to produce a developmentally competent egg and embryo, it is not surprising that maternal gene products are not only essential for normal embryonic development but also for fertility. This review provides an overview of fundamental aspects of oocyte and early embryonic development and the interference and genetic approaches that have provided access to maternally regulated aspects of vertebrate development. Some of the pathways and molecules highlighted in this review, in particular, Bmps, Wnts, small GTPases, cytoskeletal components, and cell cycle regulators, are well known and are essential regulators of multiple aspects of animal development, including oogenesis, early embryogenesis, organogenesis, and reproductive fitness of the adult animal. Specific examples of developmental processes under maternal control and the essential proteins will be explored in each chapter, and where known conserved aspects or divergent roles for these maternal regulators of early vertebrate development will be discussed throughout this review.

KEYWORDS

maternal-effect genes, maternal-zygotic transition, meiosis, maturation, cleavage, mitosis, zygotic genome activation, embryonic genome activation, axis formation, germline, fertility, oocyte, oogenesis

Contents

Introduction

"My mother made me do it!" Most of us have uttered this phrase at least once. It is commonly used in the years of our life when we are transitioning from dependent child to independence, have our own ideas, and are trying to take control of our own decisions for better or for worse. Use of the phrase was generally induced anytime we felt that our will was suppressed by our mother's will. Teenagers know all about maternal control and how it affects their phenotype, for example, how their social development is impaired when they cannot wear skinny jeans, low cut tops, or open Facebook accounts. Teenagers also have their own mechanisms to modulate the output of the maternal program. For example, the outfit that technically was not worn to school, but was instead stuffed into a backpack—the picture day photos do not lie. All psychology aside, none of us complain about or probably even think about the time during our development when our mother truly was in control, the period between production and fertilization of our egg and activation of our own zygotic genome. During this time, we dutifully followed our mothers' instructions; never mind that we did not have a brain when we were fertilized eggs or cleaving embryos. This review focuses on maternal regulation of early vertebrate development.

Multicellular organisms develop from a single fertilized cell. This cell is endowed with the potential to generate more cells that will interact with one another through direct contact and communicate through highly conserved signaling pathways to build an embryo with defined axes and organ systems. The embryo will continue to develop and grow into a mature adult capable of producing the cells necessary to form the next generation. In animals, this cellular interplay involves the formation of the gametes within the gonads of developing animals. Males of a species produce sperm, while females produce the larger of the gametes, the oocytes, and hermaphrodites produce both sperm and oocytes. In each sex, sperm or egg production depends on interactions and signaling between the somatic and germ line derived cells that comprise the gonad. Both cell types are ultimately crucial for gamete production, and thus for restoring the diploid DNA content through fertilization and to ensure optimal development of the resulting embryo. This review will focus on the contribution of the maternal gamete, the oocyte, which develops as the egg; when fertilized, this cell has the capacity to form all the cells of the embryo.

In addition to supplying half the DNA, eggs are full of nutrients, which in animals with large yolky eggs support the needs of the embryo until it can acquire food on its own. Eggs of all animals also contain mRNAs and proteins that are supplied to or deposited in the egg as it develops during oogenesis. These maternal gene products regulate meiosis, oocyte development, and early development of the embryo including fertilization, transitions between meiotic and mitotic cell cycles, and the switch from utilization of mRNAs and proteins provided by the mother to the embryo's own gene products during zygotic genome activation. The signals and regulation of meiosis will not be discussed in detail. However, many stages and aspects of oocyte development and egg production are conserved among vertebrates and will be touched upon. Here, each chapter emphasizes a maternally regulated process and the essential maternal genes required for that aspect of vertebrate development.

MATERNAL-EFFECT GENES

Maternal genes are those genes whose products, RNA or protein, are produced or deposited in the oocyte or are present in the fertilized egg or embryo before expression of zygotic genes is initiated. Genes whose RNA and protein products are only produced after zygotic genome activation, at or after the maternal to zygotic transition stage, are zygotic genes. Microarray expression studies have revealed numerous genes that are maternally expressed in vertebrates (Dworkin and Dworkin-Rastl, 1990; Schultz and Heyner, 1992; Wassarman and Kinloch, 1992; Temeles et al., 1994; Zimmermann and Schultz, 1994; Nothias et al., 1995; Wiekowski et al., 1997; Pelegri, 2003; Mager et al., 2006; White and Heasman, 2008). While this expression information is suggestive of function and provides comparative insight into which genes are exclusively maternal or zygotic, only functional investigation of maternally expressed genes can tell us which genes have essential maternal functions during development. It is worth noting that there are also examples of genes which are not maternally expressed but whose zygotic transcripts begin to be detected at stages prior to global activation of the zygotic genome, some of which regulate clearance of maternal mRNAs (Yang et al., 2002; Leung et al., 2003; Zhang et al., 2003; Lund et al., 2009). The mechanisms mediating the selective activation of genes prior to zygotic genome activation and in the context of global repression are intriguing and not well understood.

INVESTIGATING MATERNAL FUNCTION

An embryo exclusively relies on maternal gene products, RNAs, and proteins for its early development until activation of its own genome. The precise developmental time period and developmental processes under maternal control when the embryo is largely transcriptionally silent vary among organisms. In some animals, such as mice, humans, and nematodes (*Caenorhabditis elegans*), only the first or first couple of cleavage cycles are accomplished before transcription of the embryonic genome is activated. Other organisms, such as *Drosophila*, *Xenopus*, and zebrafish, rely on maternal RNAs and proteins for a more prolonged developmental period that includes several additional rounds of cleavage cycles. This dependence on maternal products persists to include regulation of early embryonic patterning, morphogenesis, and even extends to developmental processes occurring well after most maternal gene products are expected to have been degraded (e.g., mitotic divisions in mouse, late gastrulation, segmentation, and pattern formation in zebrafish and *Xenopus*). Specific examples of persisting maternal function will be discussed throughout this review and in the chapters on persisting maternal function.

What is a maternal effect? Recessive mutations disrupting zygotic gene function produce visible phenotypes only when the embryo is of homozygous mutant genotype and, thus, has two copies of the mutated or abnormal gene. Embryos with only one copy of the mutant allele are heterozygous and so appear normal. If a zygotically required gene is also maternally expressed, the maternal gene products can weaken the developmental consequences even when zygotic gene function is absent in the genotypically homozygous mutant progeny. In such cases, the embryos will have a milder defect than would otherwise be predicted based on gene expression pattern or abundance because the maternal product perdures and compensates for impaired zygotic gene function. Several examples of maternal compensation will be discussed further in the later chapters.

In contrast to zygotic recessive mutations, where visible phenotypes are observed only in homozygous individuals (Figure 1), homozygous mutant individuals for genes with strict maternal functions (i.e., no zygotic requirement) appear morphologically normal. The normal appearance of the embryos is due to the normal gene function contributed by their heterozygous mother (Figure 1). For genes with strict maternal function, mutant males are also of normal phenotype and, in the majority of cases, are fertile carriers of the defective gene. Mutant females are also morphologically normal until they reach reproductive maturity. In this review, a female is not sterile if she is able to produce mature eggs that can be fertilized, although the female is technically sterile in that her progeny are not competent to undergo normal development. All of the eggs or offspring produced by these mutant females will show developmental defects, even if they are genotypically heterozygous (Figure 1). For these progeny, loss of the maternally contributed product or failure of the mutant mother to provide normal maternal gene function is thus in effect dominant; these mutations are called maternal-effect lethals.

When we consider that maternal gene products are produced during oogenesis and that development of an embryo requires the product of meiosis, a developmentally competent egg, it is reasonable to consider genes essential for oocyte development among genes with maternal function and thus to include oogenesis as a maternal-effect process. However, the focus of this review will be largely limited to genes, or mutant alleles of those genes that are not essential for egg production, but instead impact patterning, fertilization, activation of the egg, and later aspects of embryonic

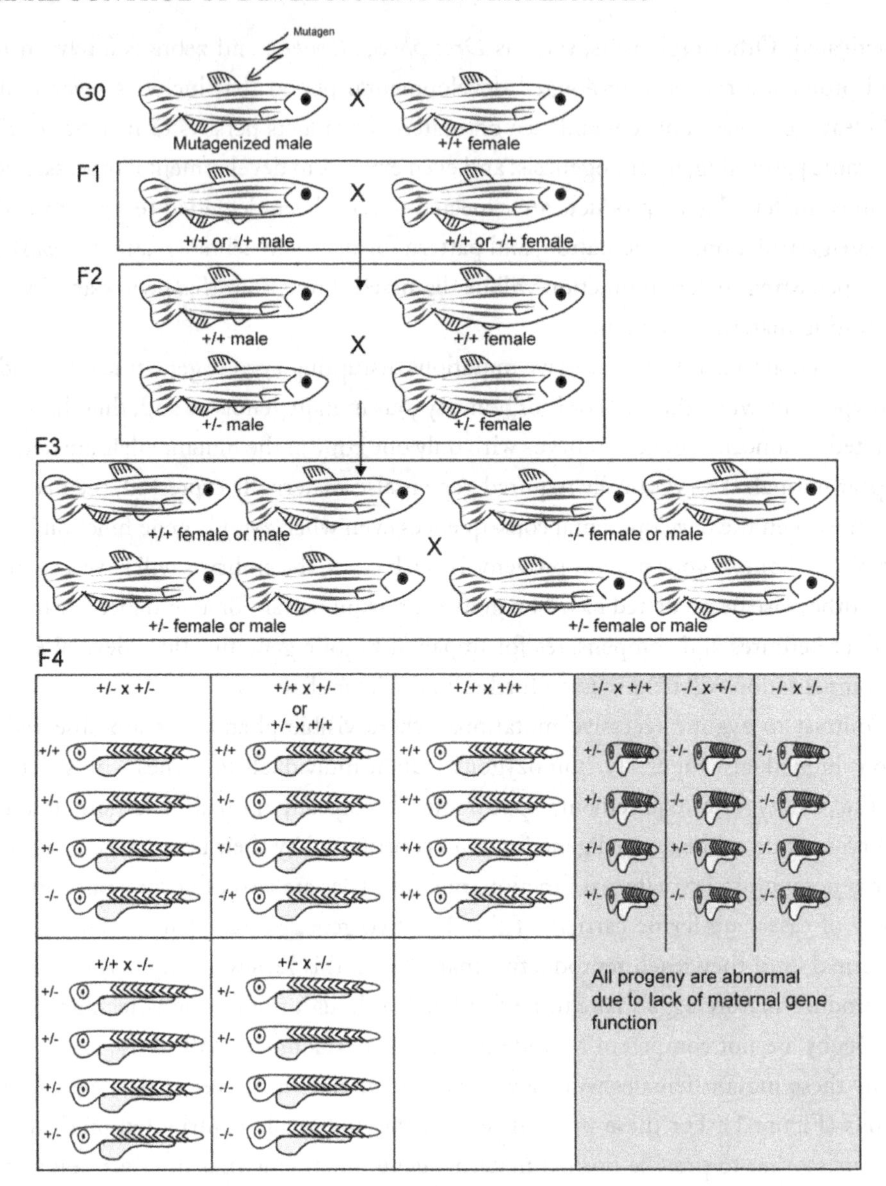

FIGURE 1: Genetic screens to uncover maternal-effect genes and maternally regulated processes. In zebrafish, males are mutagenized (G0). Mutagenized males are crossed with females of the same genetic background to generate F1 carries of newly introduced mutations. Intercrossing F1 siblings generates F2 progeny among which half carry the mutation. Twenty-five percent of the F3 progeny produced from a cross of two heterozygous carriers are mutant, but these mutants appear morphologically normal for maternal-effect genes. Mutant females are identifiable by examining their F4 progeny. The progeny of mutant females are phenotypically abnormal even if they are of wild-type genotype owing to failure of their mother to supply the maternally required gene function.

development. While mutations compromising uterine competence in mammals, such as *p53* and its target *leukemia inhibitory factor*, play key roles in maternal reproduction, these molecules will not be discussed in detail in this review (Hu et al., 2008; Hu et al., 2007; Kang et al., 2009). Reproductive defects will be included among the maternal-effect genes if they are localized to the germline and do not preclude formation of a fertilization competent egg.

TARGETING MATERNAL-EFFECT GENES

The source of maternal RNAs and proteins include molecules produced by the developing oocyte as well as somatic cells, which produce products that are later imported into the oocyte. In *Drosophila*, *C. elegans*, *Xenopus*, fish, and mammals proteins or their precursors are imported via the maternal bloodstream or from the somatic cells of the follicle. These gene products together with the RNAs and proteins produced by the oocyte constitute the maternal pool. The abundance of maternal proteins and transcripts in the developing oocytes within the vertebrate ovary makes maternal products difficult targets for interference because manual injection of interfering forms of the gene products, either plasmid DNA, messenger mRNA, or antibodies against the protein of interest are difficult to deliver or cannot be delivered in time to effectively block gene function before it is required.

Recently, several methods have made it possible to target or interfere with maternal gene function in model organisms. These approaches and some advantages and limitations of each are discussed briefly and are summarized in Table 1. Due to variability in the effectiveness of each approach in distinct model systems and stage-specific limitations due to oocyte size, number, or accessibility, it remains necessary to develop and utilize multiple approaches tailored to a specific model system, developmental process, or stage to study the maternal function of a candidate gene.

GENETICS/MUTAGENESIS

In contrast to the limitations of transient depletion, mutants, embryos with genetic lesions caused by chemical mutagens, radiation, retroviral insertion, or zinc finger nucleases disrupting genes required maternally for development of oocytes or early embryos allow perturbation of gene function at the earliest stage when it is required without causing off-target effects. Such maternal-effect mutants have been discovered in large-scale mutagenesis screens in model organisms including *Drosophila melanogaster* (flies), *C. elegans* (worms), *Danio rerio* (fish) and made by homologous recombination to disrupt candidate genes in *Mus musculus* (mouse) (Table 1) reviewed by St Johnston and Nusslein-Volhard (1992), Kemphues and Strome (1997), Pelegri (2003), Pelegri and Mullins (2004), Abrams and Mullins (2009), and Lindeman and Pelegri (2009). Though maternal-effect screens have only recently been carried out in vertebrates, such mutants have already identified novel maternal-effect functions for genes with essential maternal roles during early embryonic development. The maternal-effect screens in zebrafish yielded the largest collection of maternal-effect mutants in vertebrates; nevertheless, only a fraction of the total expected were identified, based on the

TABLE 1: Methods used to interfere with maternal gene function in model systems.

METHOD OF INTERFERENCE	DESCRIPTION OF METHOD	ATTRIBUTES	LIMITATIONS	ORGANISMS
RNA interference (RNAi) shRNAs	Females or oocytes are injected, soaked, or fed double stranded RNA. Alternatively transgenesis is used to introduce a transgene that will produce the shRNA	—Can be specific —Effective/potent —Amenable to High-throughput/broad coverage of genomes —Germline transmission in *C. elegans*	—Variable efficacy —Tissue-specific responses (some tissues are not sensitive to RNAi) —Some off-target phenotypes —May be partial/incomplete knock-down	*C. elegans* *M. musculus* *D. melanogaster*
siRNA	Complimentary single- stranded approximately 19-mers are produced, and transfected or microinjected into cells.	—Interfere with specific gene —Amenable to high-throughput/broad coverage of genomes	—Variable efficacy —Tissue-specific responses —Some nonspecific effects —Transient —May be partial/incomplete knock-down	*Xenopus laevis*
Morpholino/antisense oligonucleotides	Antisense DNA oligonucleotides targeting the start codon or splice sites. Introduced by injection or transfection.	—Can be more specific than dominant negative —Rapid —Relatively inexpensive	—Potential off target effects —Experimentally demanding —Transient —May be partial/incomplete knock-down	*Xenopus laevis* *D. rerio*

Dominant negative	Form of the gene product that can compete competitively with the wild-type form of the gene product or otherwise interfere with the wild-type activity. Usually introduced by microinjection.	—Rapid —Relatively inexpensive —Can block existing maternal proteins	—Can be nonspecific. May inhibit multiple pathways, proteins —Nonuniform effects/depend on translation efficiency and stability of protein	*D. melanogaster* *Xenopus laevis* *D. rerio*
Large-scale Mutagenesis Maternal-effect screens	Chemical (ethyl nitrosurea, ENU, ethyl methane-sulfonate, EMS) or gamma radiation	—Random/unbiased —Usually single gene is impacted. Nonlinked alleles can be Segregated from the phenotype of interest. —Transmitted through germline —Potential to identify null or hypomorphic alleles —Germline transmission	—Random —Time consuming. Several generations (F4) are required to see maternal effects —Essential zygotic functions can limit access to maternal genes —May not get mull alleles and some alleles are neomorphic	*C. elegans* *D. melanogaster* *D. rerio*
TILLING Target induced lethal lesions in genomes	Mutagenize as described above for large-scale mutagenesis. Use celI digest and sequencing to identify mutations in a gene of interest	—Can identify null mutations in a gene of interest without screening for phenotypes —Know that you have a mutation in your gene of interest —Germline transmission	—Limited by the exon size of the target gene, although advances in sequencing technology may alleviate this limitation.	*D. rerio* *C. elegans*

TABLE 1: (*continued*)

METHOD OF INTERFERENCE	DESCRIPTION OF METHOD	ATTRIBUTES	LIMITATIONS	ORGANISMS
Site-directed mutagenesis (recombineering)	Homologous recombination based replacement of coding sequence	—Specific —Efficient —Relatively rapid —Germline transmission	—Essential zygotic functions can limit access to maternal genes.	*M. musculus*
Germline replacement	The host germ line is replaced with germ cells derived from a mutant donor animal	Allows access to genes with both maternal and zygotic functions	Tedious and demanding experiments	*D. rerio*
Germline clones	Homologous recombination is introduced in germline cells	Allows access to genes with both maternal and zygotic functions	Mosaic	*D. melanogaster*
Gene trap	Gene trap vectors typically consisting of a reporter gene lacking a promotor and a selectable marker are flanked by splice acceptor sequence and a polyadenylation sequence. Reporter gene expression is driven by the promoter of the trapped gene. The premature termination site within the gene trap vector disrupts the endogenous gene.	Allows access to genes with both maternal and zygotic functions	Insertion does not always disrupt gene function	*M. musculus* *D. rerio*

number of genomes screened and estimates from previous large-scale zygotic screens (Mullins et al., 1994; Solnica-Krezel et al., 1994; Driever et al., 1996). Specifically, after screening 600 genomes, the vast majority of zygotic mutations were represented by single alleles, and less than one-half of the expected genes were identified. The zebrafish maternal-effect screens combined to date approach these numbers; however, these screens select against maternal genes that also have essential zygotic functions (Dosch et al., 2004; Luschnig et al., 2004; Pelegri et al., 2004).

GERMLINE REPLACEMENT: BYPASSING ESSENTIAL ZYGOTIC FUNCTIONS

Germline replacement has been a successful approach to investigate the maternal function of genes with essential zygotic requirements during zebrafish development (Ciruna et al., 2002). Briefly, this approach involves elimination of the host germline using a morpholino (discussed further in the next section) to prevent production of a gene product essential for survival of the germline stem cells (Ciruna et al., 2002). The germline of the host embryo is replaced by transplantation of labeled germline stem cells of a mutant donor. The resulting fish have wild-type somatic composition and mutant germline tissues; thus, allowing examination of the maternal function of the mutated gene. This approach is very effective, but is also labor intensive. Therefore, alternative approaches have been pursued to further access the essential maternal genes and to improve our understanding on the molecular and cellular basis of maternally regulated aspects of vertebrate development.

MORPHOLINOS AND OLIGO RNaseH

Advanced or maturation competent stage oocytes can be harvested from the ovaries of some vertebrate females, including mice and frogs, and injected with an interfering molecule, for example, a dominant negative form of a protein, morpholinos (antisense oligonucleotides designed to either block splicing of messages or translation of proteins from mRNAs) (Coonrod et al., 2001; Kanzler et al., 2003), or by injecting stabilized versions of antisense DNA oligonucleotides to induce RNaseH-mediated depletion of maternal mRNAs (Table 1) (Coonrod et al., 2001; Mir and Heasman, 2008). Once injected, the oocytes are induced to undergo maturation. This is accomplished either by returning the oocytes to a surrogate mother or by inducing maturation by supplying hormones to cells in culture. Finally, the manipulated oocytes are fertilized, via natural or *in vitro* approaches, and the resulting embryos are examined for developmental abnormalities. These methods work well in late-stage oocytes of *Xenopus* and mouse model systems (for more detailed methodology, the reader is referred to Colledge et al., 1994 and Mir and Heasman, 2008). Late-stage vertebrate oocytes are large and relatively easy to manipulate compared with primary oocytes. Similar approaches to conduct an interference study in which early-stage oocytes are manipulated, cultured to maturation-competent stages, fertilized to examine the consequences to early development of the embryo have not been feasible thus far. Despite the limited developmental potential of manipulated oocytes, these approaches are a useful means to obtain a snapshot of how a gene contributes to a particular stage of oocyte development. For example, mRNA, morpholino, antibodies, or toxin can be injected into oocytes to examine the effect on mRNA localization in early- or late-stage oocytes and the ability of the oocyte to undergo maturation or to eliminate a gene product in oocytes to examine its role in the early development of the embryo.

In this review, examples where morpholinos or oligonucleotide-mediated depletion have been applied to block the translation or stability (e.g., oligonucleotide induced RNaseH-mediated depletion) of maternally supplied mRNAs in the early embryos of zebrafish, *Xenopus*, and mouse are highlighted. Unlike the situation in *Xenopus* and mouse, oligonucleotide interference technology has only been applied to the study of zebrafish oocytes in a limited context in which the morpholino was used to block translation of an injected mRNA (Bontems et al., 2009). In principle, as long as the morpholino can be delivered before the protein is produced, it can be an effective approach to transiently study a gene's maternal function.

RNAi

The RNA interference pathway, which regulates gene activity and acts to protect cells against pathogens, has been successfully co-opted by researchers as a strategy for depleting gene function by interfering with maternal messages in oocytes (Svoboda et al., 2000; Svoboda et al., 2001). Double-stranded RNA, dsRNA, targeting a gene of interest, is introduced to the animal by feeding, by soaking, or introducing hairpin-producing transgenes or viruses by direct microinjection or transfection. The consequence is degradation or silencing of the target gene product, and in some organisms, such as the worm, this silencing is transmitted through the germline (Fire et al., 1998). RNAi is not very effective in translationally quiescent cells; however, when transcribed messages become translationally active during oocyte maturation in late stages of oogenesis, RNAi is an effective means to interfere with gene function (Svoboda et al., 2000; Svoboda et al., 2001; Kennerdell et al., 2002). A clear advantage of the RNAi approach is that the animal and the gonad are not physically manipulated or removed from the animal during the depletion, thus limiting unintended and undesired nonspecific perturbations due to physical manipulation.

RNAi hairpin technology has recently been reported to be effective in zebrafish (Dong et al., 2009). The strategy takes advantage of the efficient production of short hairpin RNAs by the miR30 transcribed region when expressed from a ubiquitous or tissue-specific promoter (Dong et al., 2009). The endogenous hairpin is replaced with a hairpin targeted to a gene of interest, which when processed will produce a heritable transgene capable of producing interfering RNAs. Although this shRNAi strategy has not yet been applied to study a maternal-effect function, when combined with tissue-specific promoters, this knockdown strategy should improve access to candidate genes with both essential zygotic and maternal functions.

Together, these molecular and genetic approaches provide access to maternally regulated processes that are conserved yet poorly understood, including features of oocyte and early embryonic development.

• • • •

Oogenesis: From Germline Stem Cells to Germline Cysts

Egg production or oocyte development begins with mitotically active germline stem cells. Mitotic divisions of germline stem cells can be self-renewing and lead to the production of more stem cells or can yield cells with the potential to develop as oocytes or cystoblasts. Cystoblast cells are found in juveniles of most animals and do not directly enter meiosis and begin developing as oocytes. Instead, these cells undergo several rounds of "amplifying" mitotic divisions. Cystoblast cell divisions are characterized by incomplete cytokinesis, which generates a group of interconnected cells (a cyst; Figure 2) (Trentini and Scanabissi, 1978; Pfannenstiel and Grunig, 1982; Storto and King, 1989; Pepling and Spradling, 1998; Pepling and Spradling, 2001; Kloc et al., 2004a; Neaves, 1971; Gondos, 1987; Saito et al., 2007; Marlow and Mullins, 2008).

In females of *Xenopus* or teleosts, including Medaka and zebrafish, all or many of the cells within the germline cysts are thought to develop as oocytes while others are eliminated by apoptosis (Kloc et al., 2004a; Saito et al., 2007; Marlow and Mullins, 2008). Therefore, in these vertebrates, the number of cystoblast divisions is not expected to impact specification of the oocyte. Indeed, the products of germline stem cell divisions can undergo distinct molecular programs either differentiating as individual oocytes apparently without forming interconnected cysts or undergoing divisions to generate cysts (Figure 2B, C) (Saito et al., 2007). Experiments combining eGFP labeled germ cells or somatic cells with live imaging identified two populations of labeled cells: individual germline cells and germline cells in clusters (Saito et al., 2007). Through these labeling studies and BrdU pulse chase experiments that follow the germ cell divisions, Saito and colleagues determined that two modes of division are utilized in juveniles of Medaka: cyst generating and individual cell-generating modes. The interconnected cysts accounted for 58% of the divisions, while the remaining divisions produced individual differentiating cells (type I; 42%) (Figure 2) (Saito et al., 2007). The observation of divisions producing individual oocytes without forming cysts (type I) indicates that, in Medaka, interconnections are not absolutely essential for oocyte specification or polarity.

Analyses of an ENU-induced mutation, *zenzai*, that causes reduced germ cell numbers in mutants provided evidence that both modes of division contribute to oocyte production and revealed

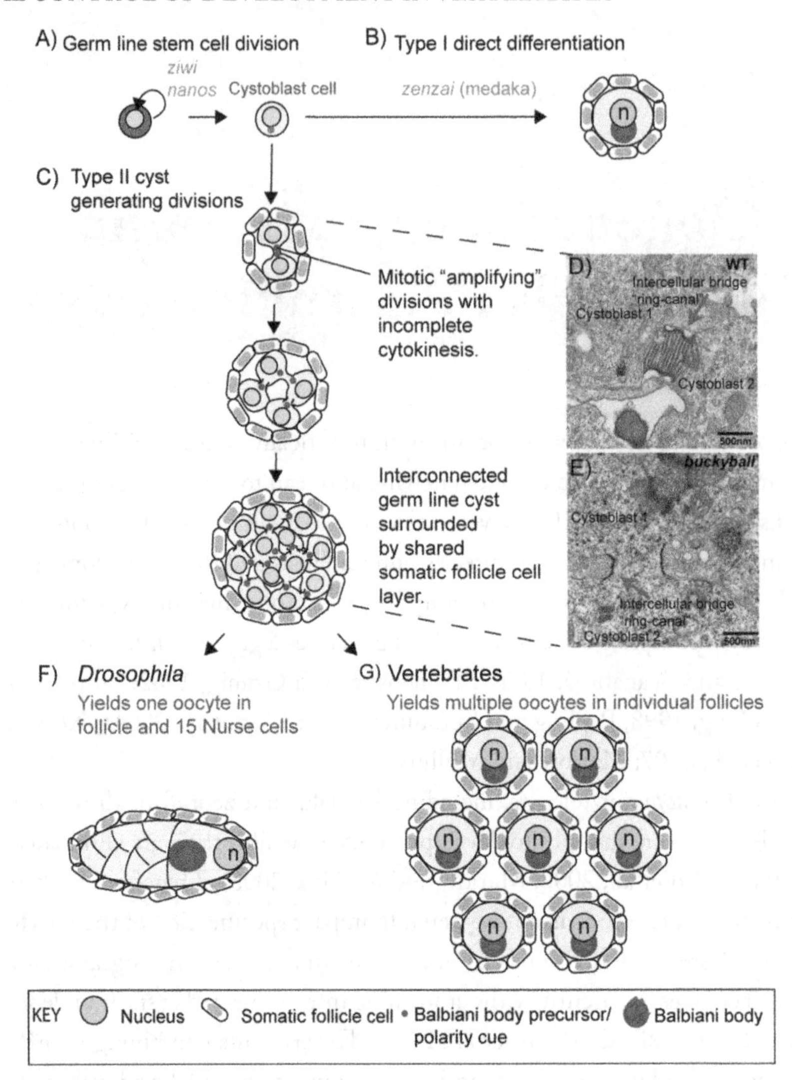

FIGURE 2: (A) Asymmetric divisions of germline stem cells produced polarized primary oocytes. Two-modes have been observed in vertebrates. (B) Type I divisions generate individual daughter cells, while (C) Type II divisions produce amplifying or cyst generating daughters. (D, E) In cysts, the cells are connected by intercellular bridges. (F) In *Drosophila* cysts generate a single oocyte, while in (G) vertebrates several oocytes are produced from each cyst. One hypothesis is that oocytes receive polarizing cues while in the interconnected cyst.

an essential role for type I divisions in maintaining the germline of Medaka adults (Saito et al., 2007). In *zenzai* mutants, cyst divisions (type II) are intact; however, germ cells are depleted in the adult as a consequence of defective (type I) germline stem cell divisions (Saito et al., 2007). The molecular mechanisms and pathways that determine whether the progeny of the germline stem cells produce a cyst of interconnected cells rather than directly differentiating to produce an individual cell are not clear. However, the *zenzai* mutant phenotype points toward control of these two modes of oocyte production by distinct molecular programs.

In the neonatal mouse, ring canals or intercellular bridges are thought to limit the number of cells that survive to later develop as individual oocytes in follicles (Pepling and Spradling, 1998; Pepling and Spradling, 2001). Treatment with phytoestrogens causes ring canals to persist and results in follicles that contain multiple oocytes rather than a single oocyte per follicle at stages after the cysts have resolved (Pepling and Spradling, 1998; Pepling and Spradling, 2001). This might suggest that turnover of ring canals or intercellular bridges play an important role in ensuring a single oocyte occupies each follicle. However, mice with mutations in *tctex/Tex14*, an essential protein for ring canal formation and male fertility in *Drosophila* and also in mouse, produce fewer oocytes rather than displaying a complete failure to produce oocytes (Caggese et al., 2001; Li et al., 2004; Greenbaum et al., 2009). If ring canals are essential for oocyte determination or production, one might expect a complete loss of oocytes. The *tctex* knockout phenotype indicates that intercellular connections are not absolutely essential for oocyte production in mice. However, it does not exclude the possibility that the synchronous cystoblast divisions observed in juveniles are important for increasing or amplifying the number of potential oocytes and thus function to promote optimal reproductive potential. It is also possible that mice, like Medaka, use multiple modes of germline stem cell divisions to generate oocytes, but whether or not this is indeed the case remains to be examined.

In vertebrates, germline cysts are prominent in juveniles and are less apparent in adult gonads (Pepling et al., 1999; Kloc et al., 2004a; Saito et al., 2007; Marlow and Mullins, 2008). Therefore, it is possible that the cyst mode is favored or only operates in the juvenile phase of development during establishment of the gonad. In mammals, females produce all of their germline stem cells before birth; thus, the cyst mode would serve as an efficient means to generate a larger pool of cells than would be possible with divisions that produce only individual cells. Cyst or "amplifying divisions" also limit the number of germline stem cell divisions, thus reducing the opportunity to accumulate mutations in this stem cell population. It is possible that the regulator of cyst type divisions is developmentally restricted and that the individual mode continues to operate in vertebrate species that produce oocytes as adults. Additional detailed analyses of germline cyst development and how it is regulated in vertebrates are needed to fully understand the contributions of cyst type and individual type divisions.

The formation of these interconnected germline cysts is a conserved feature among many animals including insects, frogs, fish, and mammals, but is best understood in *Drosophila* (Figure 2) (Filosa and Taddei, 1976). The interconnections or intercellular bridges between sister cells form from an arrested cleavage furrow during the mitotic divisions, which further develops as an actin-rich structure (Grieder et al., 2000; Snapp et al., 2004; Roper and Brown, 2004; Lighthouse et al., 2008). In *Drosophila*, incomplete divisions are thought to facilitate synchronous development of cells within the cyst (Deng and Lin, 2001; Snapp et al., 2004). The intercellular bridges also facilitate direct communication between sister cells through exchange of material via the interconnected cytoplasm and a structure called the fusome (Snapp et al., 2004; Lighthouse et al., 2008).

The fusome is a highly branched structure derived from mitotic spindle debris accumulated with each cystocyte division, such that new fusome material fuses with material from the previous divisions (Lin, Yue et al., 1994; Lin and Spradling, 1995; Deng and Lin, 1997; Grieder et al., 2000; Riparbelli et al., 2004). The fusome is composed of a specialized membrane skeleton that connects cyst cells through the ring canals, or intercellular bridges, to facilitate cytoskeletal organization, communication, intercellular transport, and synchrony during cyst development (Suter et al., 1989; Suter and Steward, 1991; Theurkauf et al., 1993; Lin Yue et al., 1994; Lin and Spradling, 1995; de Cuevas and Spradling, 1998; Snapp et al., 2004; Lighthouse et al., 2008). The fusome is present at only one mitotic pole of each cystocyte; thus, the divisions of cystocytes are inherently unequal (Lin et al., 1994). The fusome is highly vesiculated and, in flies, consists of endoplasmic reticulum (ER), actin, spectrins, adducin homologs, and key protein regulators of GSC/cystoblast differentiation (Lin et al., 1994; Snapp et al., 2004; Kai et al., 2005; Xi et al., 2005; Mukai et al., 2006; Chen and McKearin, 2005; Morris et al., 2005; Szakmary et al., 2005; Leon and McKearin, 1999; Lee et al., 2000; Ohlstein et al., 2000; Page et al., 2000; Parisi et al., 2001; Chen and McKearin, 2003; Chen and McKearin, 2003; Niki and Mahowald, 2003; Shivdasani and Ingham, 2003; Casanueva and Ferguson, 2004; Gilboa and Lehmann, 2004; Schulz, Kiger et al., 2004; McKearin and Christerson, 1994; Wei, Oliver et al., 1994; McKearin and Ohlstein, 1995; Gonczy, Matunis et al., 1997; Ohlstein and McKearin, 1997; Fuller, 1998; Kim-Ha et al., 1999; Lavoie et al., 1999). Similarities between the germline cyst architecture between *Xenopus* and *Drosophila* based on the number of cystocytes produced per cyst and the polarized localization of fusome-associated markers, including spectrin, adducin, and centrin within the germline cyst, indicate that development of this structure may be conserved (Kloc et al., 2004).

Genetic studies of *Drosophila* have demonstrated a role for the ring canals in maintaining synchrony, in regulating the number of cystoblast cell division cycles, and in deposition of maternal components (Storto and King, 1989; Lin et al., 1994; Ohlstein and McKearin, 1997; de Cuevas and Spradling, 1998; Lavoie et al., 1999; Leon and McKearin, 1999; Liu et al., 1999; Matuszewski et al., 1999; Grieder et al., 2000; Lilly et al., 2000; McKearin and Ohlstein, 1995; Deng and Lin,

1997; Cox and Spradling, 2003; Bolivar et al., 2001; Cox et al., 2001; Cox et al., 2001; Deng and Lin, 2001; Vaccari and Ephrussi, 2002; Mathe et al., 2003; Riparbelli et al., 2004; Roper and Brown, 2004; Snapp et al., 2004; Djagaeva et al., 2005; Bogard et al., 2007; Roper, 2007; Lighthouse et al., 2008). Although germline cysts in *Drosophila* contain 16 cells, only one of these 16 cells will develop as an oocyte (Figure 2). The other 15 cells develop as polyploid nurse cells. Nurse cells produce RNA, and protein, and transfer or "dump" these molecules together with their mitochondria and other organelles into the developing oocyte. To ensure that the oocyte is specified properly and has an appropriate amount of material for subsequent development, it is crucial that the number of cystoblast divisions is precise. Indeed, mutations that disrupt ring canal growth, development, or actin accumulation, lead to smaller ring canals and inefficient transfer of maternal components to the oocyte (Yue and Spradling, 1992; Jordan and Karess, 1997; Robinson et al., 1997; Dodson et al., 1998; Roulier et al., 1998; Li, Serr et al., 1999; Gorjanacz et al., 2002; Tan et al., 2003; Djagaeva et al., 2005; Gorjanacz et al., 2006). The resulting eggs are smaller, and embryos when they develop have defects in morphogenesis. It is also essential to form the appropriate number of ring canals. Excess or too few cystoblast divisions result in ovarian tumor phenotypes, lead to failure to specify the oocyte, and, thus, compromise fertility and developmental potential (McKearin and Ohlstein, 1995; Deng and Lin, 1997; Lavoie et al., 1999; Ohlstein et al., 2000; Parisi et al., 2001; Narbonne-Reveau et al., 2006).

• • • •

Oocyte Polarity and the Embryonic Axes: The Balbiani Body, an Ancient Oocyte Asymmetry

The divisions of the germline stem cell are asymmetric in that they generate a renewing cell and a cell with the capacity to differentiate as an oocyte. Thereafter, the asymmetric meiotic divisions of oocytes produce one large cell, the egg, and two small polar bodies. In addition to asymmetric divisions, asymmetries are present in the form of localized cellular structures, organelles (mitochondria, endoplasmic reticulum, the oocyte nucleus), and also in the distribution of proteins and RNAs within the oocyte. All or a subset of these asymmetries are present in nearly all animals (Figure 3). For example, in most animals, the nucleus is in a central position within the early oocyte, but as oogenesis progresses, the oocyte nucleus moves to the oocyte periphery, or cortex (Figure 3). The side of the cell where the oocyte nucleus is localized in late-stage vertebrate oocytes is generally defined as the animal pole (Figure 3). Earlier asymmetries in oocytes are also shared among animals.

The Balbiani body was first noted more than 100 years ago and since has been identified in primary oocytes of all animals that have been examined including mammals although the constituents vary among species (Figure 4). The Balbiani body is a transient collection of organelles, inclusions, and molecules that assembles adjacent to the nucleus of primary oocytes (Figures 3 and 4) (Guraya, 1979). In frog, fly, fish, mouse, and other mammals, including humans, the Balbiani body is asymmetrically positioned and is composed of endoplasmic reticulum, mitochondria, Golgi, and proteins (Pepling et al., 2007; Billett and Adam, 1976; Bukovsky et al., 2004; Kloc et al., 2008; de Smedt et al., 2000; Kloc et al., 2004a). In addition to these organelles, RNAs, in particular, those encoding germ plasm and patterning molecules are sequestered in the Balbiani body of zebrafish and frogs (Kloc et al., 2004a; Wilk et al., 2005). In zebrafish and in frogs, the Balbiani body functions as an integral component of a vegetal transport pathway thought to entrap mRNAs and other gene products necessary for germ cell formation and patterning of the embryo. First, these molecules colocalize with the endoplasmic reticulum, and later, they are positioned at the vegetal oocyte cortex. Thus, the Balbiani body functions to segregate these germ plasm and vegetal patterning molecules within the oocyte (reviewed by Kloc et al., 2004a).

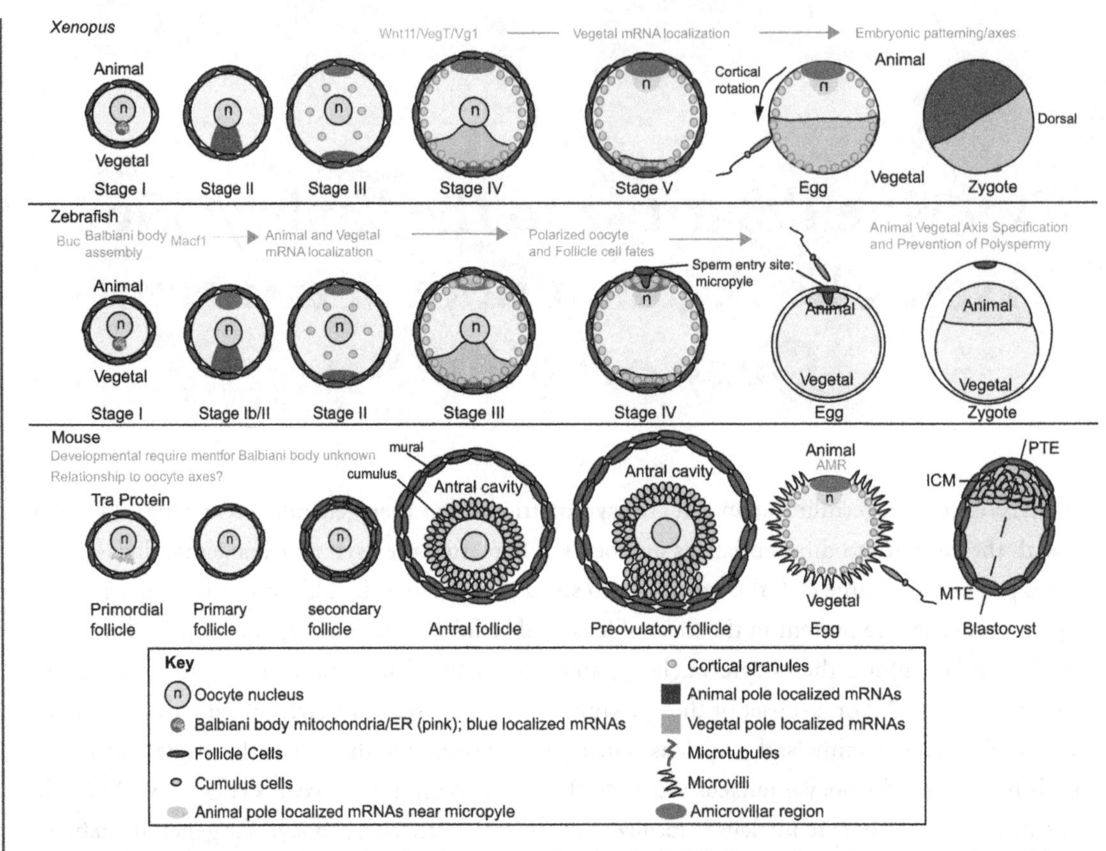

FIGURE 3: Schematics depicting development of individual follicles within the ovaries of model vertebrates. In each case, the follicle consists of a germline-derived oocyte and somatic "follicle cells." The follicles are simple at early stages, but become increasingly complex as the follicle develops. The follicles of *Xenopus* and zebrafish are considerably less complex than the mammalian follicle, although distinct granulosa and theca cells. In all of these systems, the oocyte and follicle cells communicate with and depend on one another for their survival and development. In zebrafish and in the mouse, the oocyte also influences cell fate within the somatic follicle cell layer. In zebrafish, a specialized cell, the micropyle forms only at the animal pole, and this fate is limited by a mechanism that requires polarity established by Bucky ball. In mouse, the somatic follicle cells closest to the oocyte become cumulus and those further away develop as mural cells. Conserved asymmetries are observed in the primary oocytes of vertebrate model systems. The Balbiani body marks the vegetal pole of the oocyte (stage I) and indicates the future vegetal pole of the egg and embryo (zygote) in *Xenopus* and zebrafish. Primary oocytes (Stage I) of mice also have Balbiani bodies to which conserved mRNA transport proteins localize, but whether this transient asymmetry in early oocytes of mice is predictive of the future animal–vegetal axis of the late-stage oocyte or egg is not known. ICM = inner cell mass; MTE = mural trophectoderm; PTE = polar trophectoderm; AMR = amicrovillar region.

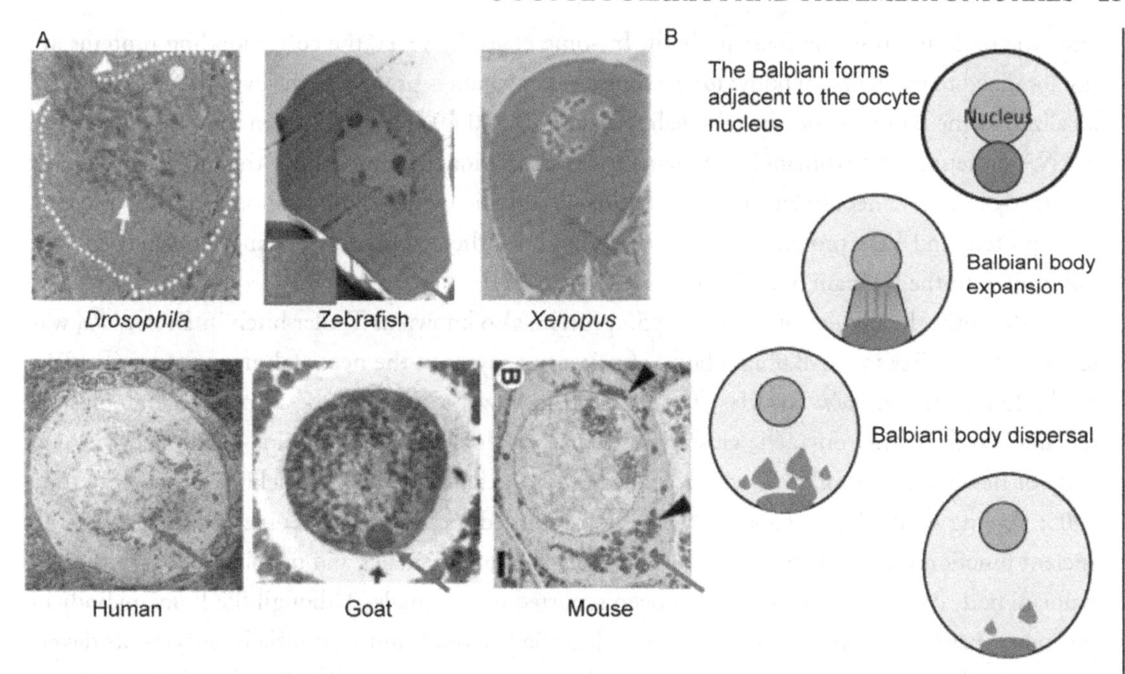

FIGURE 4: The Balbiani body is a conserved asymmetric structure in the primary oocytes from insects to mammals. (A) EM images of Balbiani bodies (pink arrows) in the primary oocytes of model organisms and humans. The Balbiani body is composed of organelles including mitochondria and ER and proteins and in some animals, mRNAs. (B) The Balbiani body is a transient structure that forms adjacent to the nucleus then undergoes expansion. In frogs and fish, the Balbiani body is a component of a pathway that delivers cargo to the vegetal cortex. The Balbiani body undergoes dispersal prior to stage II of oogenesis. Balbiani bodies from: Human; Goat: De Smedt et al., 2000. Mouse: Pepling and Spradling, 2001 *Drosophila:* Cox and Spradling, 2003; *Xenopus:* Heasman and Wylie, 1984; zebrafish: Marlow and Mullins, 2008.

Balbiani bodies share many features with P-bodies and stress bodies of other cell types. P bodies and stress bodies are distinct cellular granules containing mRNAs and enzymes that mediate mRNA turnover or storage. Hence, it is reasonable to speculate that the role of the Balbiani body might be to protect messages from being degraded and/or to prevent expression or activation of patterning molecules before they are needed. This also provides a mechanism to spatially restrict their activity. Consistent with these models, mediators of translational activation or repression are among the molecules sequestered in the Balbiani body of early oocytes (Collier et al., 2005; Marnef et al., 2009). Notably, the maternal mRNAs required at later stages to establish the germline of the embryo are localized to the Balbiani body. In some cases, however, the corresponding proteins are not localized to the Balbiani body: for example, Vasa protein is present in the cytoplasm rather than localized in the Balbiani body in zebrafish (Knaut et al., 2000). Thus, at least in the case of *vasa*, the mRNA present in the Balbiani body seems to be translationally repressed or the protein is actively excluded or not retained. In fetal ovaries of humans and mice, where Balbiani bodies are also present, Dazl protein and Vasa protein are detected throughout the cytoplasm of primary oocytes and are not limited to the Balbiani body (Anderson et al., 2007).

Recently, the highly conserved Rap55 protein, also known as Trailer hitch in *Drosophila*, was reported to localize to the Balbiani body of primary oocytes in the neonatal mouse ovary (Pepling et al., 2007). In *Drosophila* ovaries, Trailer hitch localizes to the fusome-associated endoplasmic reticulum and to other organelle clusters in the oocyte, including the anterior-positioned Balbiani body of newly formed follicles, where Trailer hitch mediates mRNA trafficking (Wilhelm et al., 2005; Pepling et al., 2007; Roper, 2007). The conserved localization of Trailer hitch supports an ancient function for the Balbiani body in the transport of organelles and possibly RNAs, although Balbiani body-localized RNAs have not been reported in mammals. Although the Balbiani body in the mouse has been proposed to function to select the healthiest mitochondria in oocytes, its developmental role remains to be demonstrated. In humans, like mouse, the Balbiani body is present as a perinuclear aggregate that includes mitochondria and other organelles and is transiently present in a spherical conformation (Figure 4) (Gondos, 1987). The developmental requirement for the Balbiani body in humans and whether it is also an indicator of the prospective animal–vegetal axis of the late-stage oocyte and egg remains to be determined.

Although the germ cell-specific mRNAs, such as, *vasa, nanos, gasz*, and *dazl*, are conserved and localize to the Balbiani body of fish and frog oocytes, Balbiani body or polarized localization of these mRNAs in the early oocytes of mammals has not been observed so far (Houston et al., 1998; Maegawa et al., 1999; Nishi et al., 1999; Houston and King, 2000; McNeilly et al., 2000; Mita and Yamashita, 2000; Yan et al., 2002; Yan et al., 2004; Braat et al., 1999; Knaut et al., 2000; Kloc et al., 2001; Kloc et al., 2002; Kosaka et al., 2007). One explanation for the absence of localized germ plasm mRNAs in mammalian oocytes can be attributed to the different modes or developmental timing of primordial germ cell specification in mammals compared with fish and frogs (reviewed by Wylie, 2000; McLaren, 2003). Specifically, in fish and frogs, the primordial germ cells are specified from maternally inherited germ plasm, whereas in the mouse, the primordial germ cells are induced during gastrulation. In addition to the germ plasm mRNAs, several molecules necessary for the development of the embryonic axes of zebrafish and *Xenopus* are localized asymmetrically along the animal–vegetal axis in oocytes. These molecules will be discussed later in the discussion of dorsal–ventral axis formation.

MOLECULAR CONTROL OF BALBIANI BODY DEVELOPMENT

Despite awareness of the Balbiani body for more than 100 years and its conserved presence in the primary oocytes of all animals examined, how Balbiani body formation is regulated at the cellular and molecular level and its essential developmental function are not understood. The conserved germline cyst architecture discussed earlier, namely, the intercellular bridges and fusome have been postulated to function as or to position a polarity cue or Balbiani body precursor materials (Figure 5) (Kloc et al., 2004; Kloc et al., 2004a; Kloc et al., 2008). However, so far, mutants are not available to functionally test whether oocyte polarity or Balbiani body "specification" is inherited from asymmetries or cues positioned within the vertebrate germline cyst (Figure 5). From studies conducted in invertebrate and vertebrate genetic systems combined, only three genes regulating Balbiani body development have been described to date and are described below.

MILTON: KINESIN ADAPTORS AND MITOCHONDRIAL ALLOCATION

Milton, a kinesin-associated adaptor protein, was identified in *Drosophila* for its function in allocation of mitochondria from the nurse cells to the Balbiani body of the oocyte (Figure 5) (Cox and Spradling, 2006). The *milton* alleles described in *Drosophila* do not impede RNA localization or disrupt axis specification (Cox and Spradling, 2003; Cox and Spradling, 2006). This finding indicates that Milton has a dedicated function in regulating mitochondria allotment to the Balbiani body, whereas transient recruitment of RNAs, such as *oskar*, to the Balbiani body is regulated independent of Milton and, apparently, also independent of mitochondria acquisition. Only a subset of the mitochondria present in vertebrate oocytes localizes within the Balbiani body, and nurse cells have not been identified in vertebrates. (Marinos and Billett, 1981; Pepling et al., 2007; Kloc et al., 2004a; Kosaka et al., 2007; Marlow and Mullins, 2008; Kloc et al., 2008; Zhang et al., 2008). Thus, mitochondrial acquisition must occur via a mechanism that does not involve allocation from the nurse cells in vertebrates.

In one study, mitochondria of *Xenopus* oocytes were labeled using a combination of activity-dependent and -independent mitochondrial markers or with dual emission mitochondrial probes that mark mitochondria and also provide an indication of membrane potential (mitochondrial activity) (Wilding et al., 2001). (Wilding et al. (2001) observed that the most highly active mitochondria were adjacent to the nucleus or within the Balbiani body. These findings suggest one model whereby the most active mitochondria could be selectively localized to the Balbiani body. It is possible that Milton acts in an analogous manner in vertebrates to allocate, or rather to select specific (e.g., the most active) mitochondria for recruitment to the Balbiani body and to regulate the number of

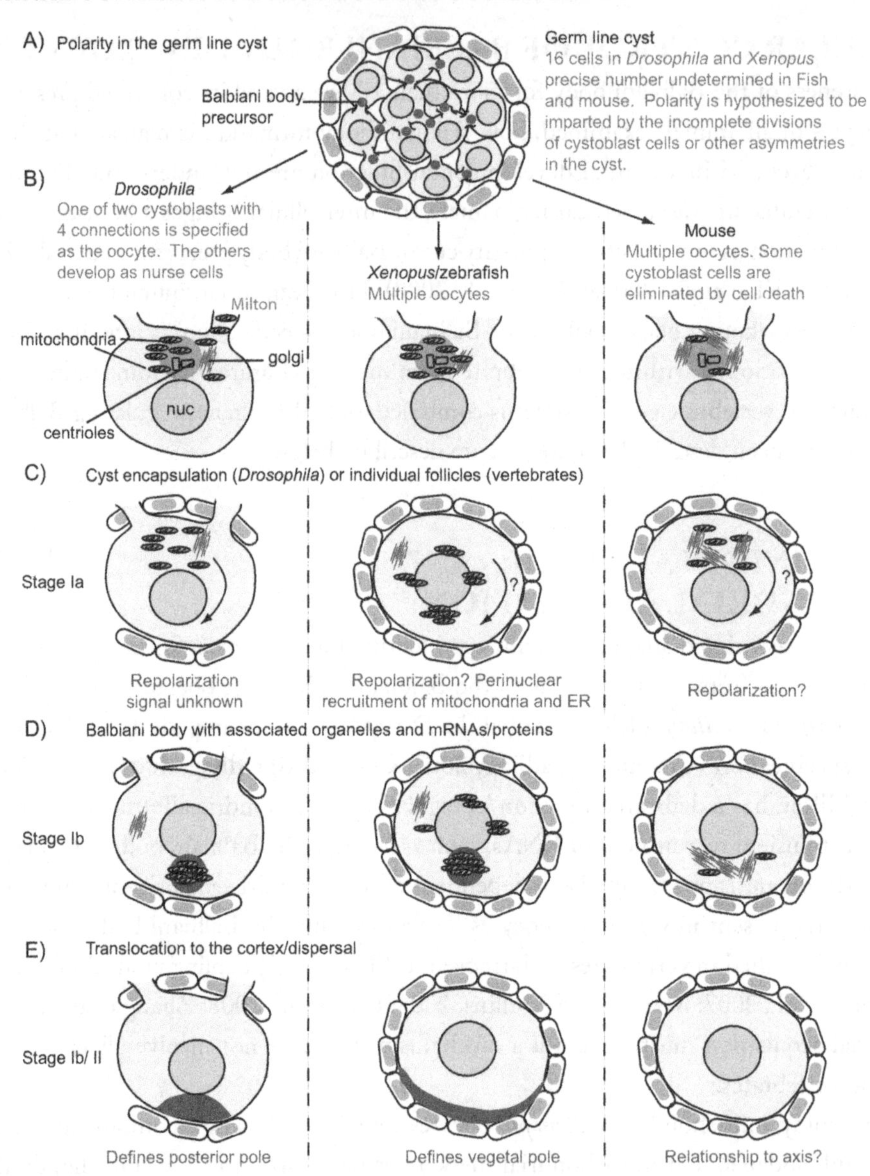

FIGURE 5: Model depicting how polarity information might be acquired from interconnected cyst to generate oocyte asymmetry.

mitochondria associated with the Balbiani body. Whether Milton or another kinesin-associated adaptor protein plays a conserved role in regulating mitochondria distribution in vertebrate oocytes remains to be determined.

BUCKY BALL: A NOVEL REGULATOR OF OOCYTE ASYMMETRY

Bucky ball was identified based on its maternal-effect egg polarity phenotype in zebrafish (Dosch et al., 2004). In the primary oocytes of zebrafish, *bucky ball* mutants mitochondria are not localized in an aggregate and no Balbiani body forms (Figure 6) (Marlow and Mullins, 2008; Bontems et al., 2009). In fact, Bucky ball protein is required to establish this earliest known oocyte asymmetry and for all examined asymmetries thereafter, including specification of the first embryonic axis to form in vertebrates, the animal–vegetal axis (Marlow and Mullins, 2008; Bontems et al., 2009).

The oocytes of frogs and fish are highly polarized as can be seen in the localization of mRNAs and proteins along the oocyte axis. The Balbiani body is the earliest known indicator of the asymmetry, and the mRNAs that localize to this structure are the first known to localize. Thus, the Balbiani body-mediated pathway is referred to as the early vegetal localization pathway (reviewed by Kloc and Etkin, 2005; King et al., 2005; Minakhina and Steward, 2005). Many of the mRNAs that localize to the Balbiani body are germ plasm components; however, a few mRNAs encoding regulators of embryonic patterning (e.g., Wnt11 in frogs, Syntabulin in zebrafish) also utilize the early Balbiani body-mediated vegetal pathway. The majority of localized mRNAs encoding molecules involved in specification or patterning of the later developing dorsal–ventral axis localize via later localization pathways. The late pathway for localization to the vegetal pole operates at stages after the Balbiani body is no longer present, but it relies on an intact cytoskeletal network thought to be organized by the earlier Balbiani body vegetal pathway (e.g., by molecules localized during or after the Balbiani body-mediated phase; reviewed by Kloc and Etkin, 2005; King et al., 2005; Minakhina and Steward, 2005). Other RNAs localize to the opposite side of the cell, at the animal pole, and only become asymmetrically distributed at later stages of oocyte development when the Balbiani body is no longer present.

Notably, in *bucky ball* mutants, even the late localizing vegetal pole RNAs, which do not transit through the Balbiani body, require *bucky ball* function. It remains to be determined whether this is due to the absence of the Balbiani body or represents an additional function of Bucky ball. Conversely, animal pole-localized mRNAs are no longer limited to the animal pole; instead, these RNAs are found around the circumference of the oocyte cortex (Marlow and Mullins, 2008; Bontems et al., 2009) (Figure 7). The expanded animal character in *bucky ball* mutants supports a model, whereby establishing vegetal pole identity breaks symmetry and regulates or alternatively indirectly influences the extent of the animal territory; however, the mechanism linking patterning of the

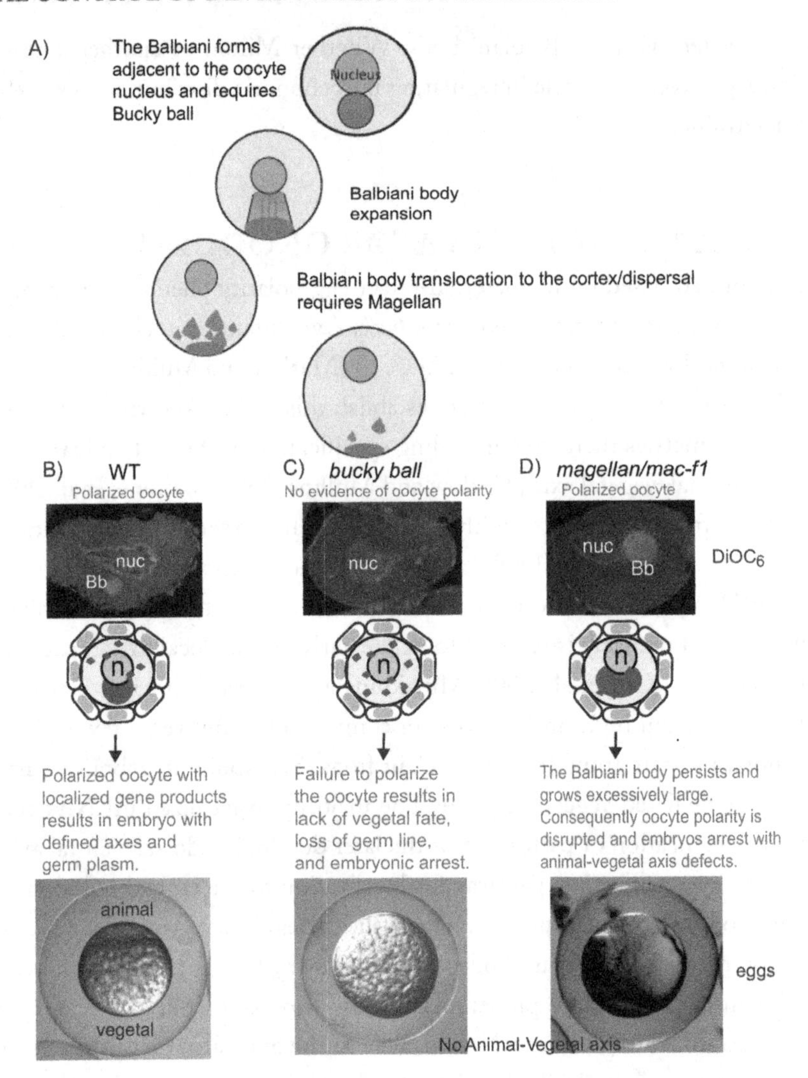

FIGURE 6: Regulators of Balbiani body dynamics in zebrafish oocytes. (A) Formation of the Balbiani body adjacent to the nucleus requires the novel protein Bucky ball, while Magellan/MacF1 (an actin-microtubule crosslinking protein) contributes to dispersal of the Balbiani body. (B) DiOC6 labeled mitochondria and ER in the wild-type Balbiani body (Bb) indicates the prospective vegetal pole and is the earliest known indicator of oocyte polarity. This polarized primary oocyte will generate an egg and embryo with defined animal–vegetal axis. (C) In *bucky ball* mutant oocytes, the Balbiani body and the vegetal pole are not defined. The resulting embryo lacks an animal–vegetal axis and is not competent for embryonic development. (D) In *magellan* mutants polarity is defined, but the Balbiani body does not disperse on time. The consequence is an egg and embryo without an apparent animal–vegetal axis.

two poles is not known. It is possible that the animal and vegetal poles are defined by a mechanism involving mutual exclusion of protein complexes at the cell cortex (e.g., similar to the mutual exclusion of Par protein complexes that polarize diverse cell types and pattern the anterior–posterior axes in *C. elegans* and *Drosophila* (Figure 7C); reviewed by Wodarz, 2002; Nance, 2005; Goldstein and Macara, 2007). Alternatively, independent signals (from within the oocyte or from somatic cells) are required to specify the vegetal and animal regions and activate mutually antagonistic programs to pattern the animal–vegetal axis (Figure 7D). So far, a "vegetalized" (i.e., expanded vegetal at the expense of animal pole) maternal-effect mutant to support active induction of animal identity has not been discovered. Thus, it is possible that the animal pole is the default state in the absence of vegetal pole specification, which, once induced, would act to inhibit or override the default animal program. Even if the animal is at a default state, evidence for active patterning of the animal pole (probably mediated in part by signals from the surrounding somatic cells) can be seen in the localization of markers such as Vg1 in *bucky ball* mutants (Figure 7B). Vg1 is normally only present in the medial animal region, but is found in multiple discrete domains at the oocyte cortex in *bucky ball* mutants, which is consistent with the presence of multiple animal poles rather than a simple expansion of animal character into vegetal regions (Figure 7B) (Marlow and Mullins, 2008).

Bucky ball is required both to assemble the Balbiani body, a marker of vegetal in zebrafish, and to regulate development of the animal–vegetal axis. A key question that remains to be addressed is whether the animal–vegetal polarity and Balbiani body defects are interdependent (i.e., defective polarity causes the Balbiani body assembly defect or *vice versa*). Alternatively, it is possible that Balbiani body assembly and axis formation are separable processes that independently require Bucky ball function. To distinguish whether defective oocyte polarity in late-stage oocytes reflects a constant requirement for Buc protein to maintain polarity or if these defects are a secondary consequence of the earlier polarity defect will require conditional rescue of *bucky ball* mutant phenotypes or conditional inactivation of the Buc protein. Genetic mutants, like the *Drosophila milton* mutants, that uncouple aspects of Balbiani body development from development of the first axis (the animal–vegetal axis in vertebrates) have yet to be discovered in vertebrates.

The molecular identity of the *bucky ball* gene does not provide insight into the mechanism by which Buc mediates oocyte polarity and Balbiani body formation. The *buc* gene encodes a novel protein of unknown function first identified in *Xenopus* as a vegetally localized transcript, *Xvelo* (Bontems et al., 2009; Claussen and Pieler, 2004). Mammals also have *bucky ball-like* genes; however, the human *buc-like* genes lack predicted open reading frames (Bontems et al., 2009). Whether the human *buc-like* genes have retained function perhaps as noncoding mRNAs and can compensate for any aspect of *bucky ball* function in zebrafish mutants remains to be investigated. The absence of identifiable functional domains in the Bucky ball protein makes it difficult to assign Bucky ball to a known signaling pathway. Currently, no other gene required to promote Balbiani body formation is

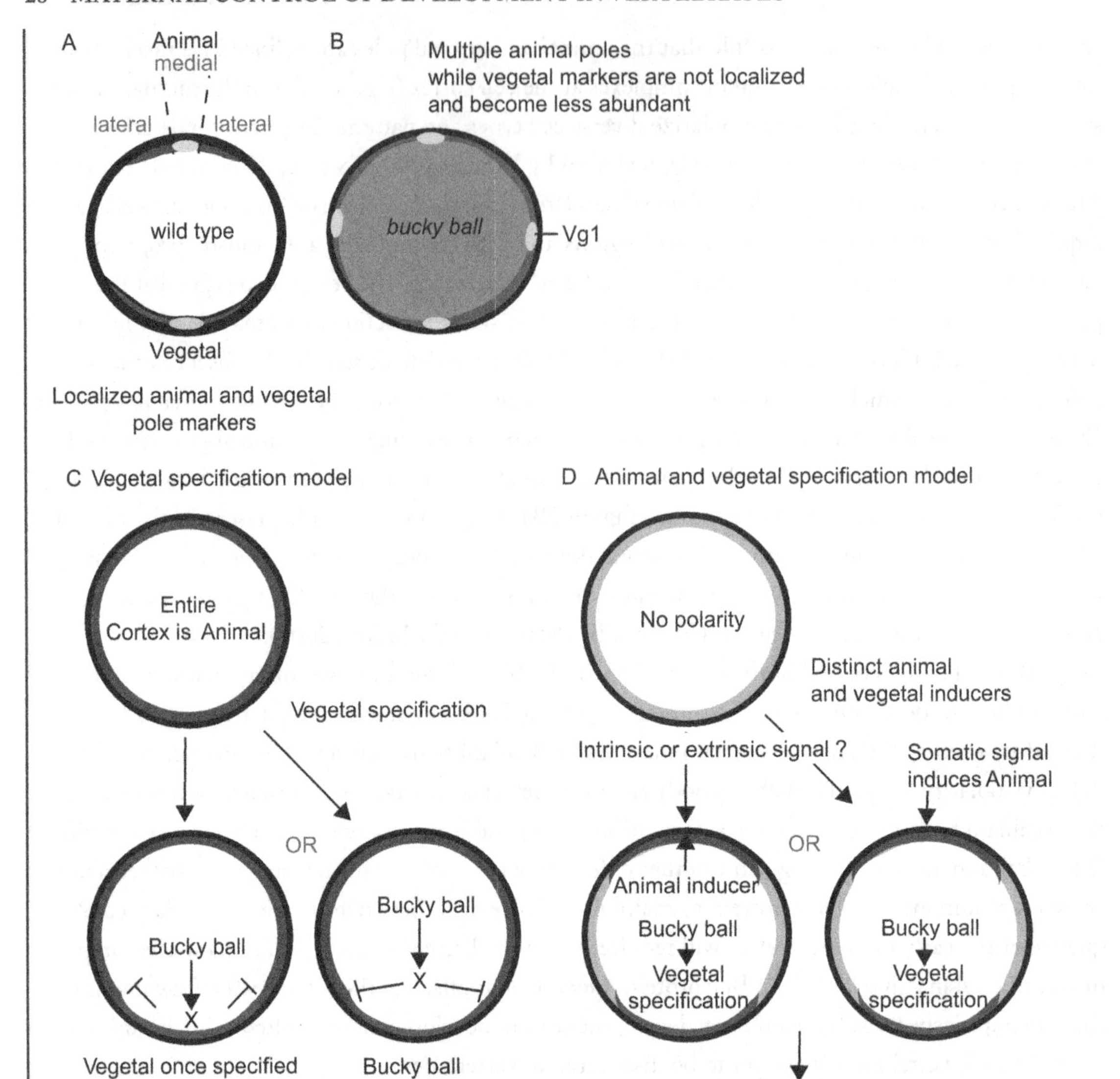
A
Animal
medial
lateral
lateral
wild type
Vegetal
Localized animal and vegetal pole markers
B
Multiple animal poles while vegetal markers are not localized and become less abundant
bucky ball
Vg1
C Vegetal specification model
Entire Cortex is Animal
Vegetal specification
OR
Bucky ball
X
Vegetal once specified represses animal
Bucky ball
X
Bucky ball mediated polarity pomotes activity to inhibit animal
D Animal and vegetal specification model
No polarity
Distinct animal and vegetal inducers
Intrinsic or extrinsic signal ?
Somatic signal induces Animal
OR
Animal inducer
Bucky ball
Vegetal specification
Bucky ball
Vegetal specification
Animal
Vegetal
Once specified animal and vegetal are mutually exclusive

known in vertebrates or invertebrate model systems. Thus, biochemical studies, additional genetic screens, and analysis of conditional knockouts are needed to identify the molecular regulators of Balbiani body formation and oocyte polarity, including novel components of the Bucky ball pathway, to fully understand how polarity is established and maintained in vertebrate oocytes.

MACF1/MAGELLAN: A ROLE FOR MICROTUBULE: ACTIN CROSS-LINKERS IN BALBIANI BODY DYNAMICS

Like *bucky ball*, *magellan* was identified through zebrafish maternal-effect screens based on the animal–vegetal polarity phenotypes of eggs (Dosch et al., 2004). In contrast to *bucky ball* mutant oocytes, which fail to form Balbiani bodies, primary oocytes of *magellan* mutants are able to recruit germ plasm mRNAs and mitochondria to establish a Balbiani body that initially resembles the Balbiani body of wild-type oocytes (Gupta et al., 2010). Similar to wild type, the Balbiani body of *magellan* mutant oocytes undergoes expansion, but then it grows abnormally large and persists in later-stage oocytes because of the failure of the Balbiani body to transit to the oocyte cortex (Figure 6) (Gupta et al., 2010). Consequently, germ plasm mRNAs are not delivered to the cortex and are effectively trapped in the persisting Balbiani body. In wild-type oocytes, acetylated (i.e., stable) microtubules are observed in the cytoplasm between the nucleus and the cortex around the circumference of primary oocytes; acetylated microtubules do not reach the cortex in *magellan* mutant oocytes, but instead surround the nucleus (Gupta et al., 2010). Thus, anchoring microtubules to the cortex may be a key step in Balbiani body-mediated cargo delivery and turnover. Magellan also maintains a centrally positioned nucleus (Gupta et al., 2010). The extent to which the large Balbiani body and displacement of the nucleus phenotypes are linked is not known. The excess growth of the Balbiani body of *magellan* mutants could preclude the nucleus from occupying a central position, or these

---

FIGURE 7: Models depicting potential mechanisms to establish animal–vegetal oocyte polarity. (A) In wild-type zebrafish oocytes markers are localized along the animal–vegetal axis. (B) In *bucky ball* mutants, vegetal markers (blue) are not localized while animal markers are found around the circumference or in multiple discrete ectopic foci. (C) In this vegetal specification model, the oocyte cortex is initially defined as animal. After vegetal is specified by a mechanism that depends on *bucky ball*, animal pole character is repressed at the cortex. Alternatively, the *bucky ball*-dependent pathway acts to inhibit animal, which then permits vegetal character to be "expressed." (D) In this model, initially the cortex is not defined as animal or vegetal. Independent signals act to promote animal and vegetal fate, which once specified will antagonize or mutually exclude one another. Additional mechanisms would also be consistent with the expression of markers in wild-type and their loss (vegetal markers) or expansion (animal markers) in *buc* mutants. Cross-section schematics.

defects may represent distinct functions of Magellan. Defective nuclear position phenotypes are not observed in *bucky ball* mutants.

Both Bucky ball and Magellan regulate animal–vegetal polarity and are essential for delivery of cargo to the vegetal cortex (Figure 6). Bucky ball promotes Balbiani body assembly, whereas Magellan is required for Balbiani body translocation and dispersal. The opposite phenotypes of zebrafish *bucky ball* and *magellan* mutants with regard to Balbiani body development raise the interesting possibility that the two genes function in the same pathway. Examination of whether *bucky ball* or *magellan* gene product abundance or localization is dependent on one another, and epistasis studies to determine the compound mutant phenotype will reveal whether these genes regulate Balbiani body development as components of a single pathway or independent pathways converging on Balbiani body dynamics.

The *magellan* mutant disrupts *microtubule actin crosslinking factor 1*, a cytoskeletal cross-linking protein previously identified as a fusome component known as *short stop* (*shot*) in *Drosophila* (Roper and Brown, 2004). Together, the molecular identity of the *magellan* gene, the defective attachment of microtubules to the cortex in *magellan* mutant oocytes, and prior known functions of Macf-1 as an organizer of the microtubule cytoskeleton indicate a mechanism whereby Magellan maintains nuclear position and facilitates translocation of the Balbiani body to the vegetal cortex via regulation of the microtubule cytoskeleton (Figure 6). *Drosophila*, a *shot/macf1* mutant, also has oocyte polarity and microtubule defects. Specifically, *shot* is necessary for polarizing the microtubules within the germline cyst, for oocyte determination, and for asymmetric accumulation of factors in *Drosophila* oocytes (Roper and Brown, 2004). The earlier polarity phenotypes in *Drosophila* oocytes compared with zebrafish *magellan* mutants may reflect the nature of the mutant alleles (e.g., it is possible that one of the multiple splice forms of *macf1* persists in zebrafish *magellan* mutants and carries out the earlier function).

In the mouse, MacF1 is essential for proper patterning of the anterior–posterior axis. Knockout of *MacF1* is zygotic lethal; homozygous mutants arrest at gastrulation due to failure to specify axial mesoderm, the absence of which ultimately leads to defective posterior body formation (Chen et al., 2006). Defective posterior body formation in the *MacF1* knockout is associated with reduced activity of the Wnt pathway, which is known to promote posterior body formation (discussed in more detail in the chapter on dorsal–ventral axis patterning) (Chen et al., 2006). The studies of Chen and colleagues support a model whereby MacF1 mediates translocation of Axin to the cell membrane where Axin is then degraded. Thus, MacF1 regulates posterior axis formation as a component of a complex that promotes Wnt activity. Whether MacF1 also has an essential maternal function in the mouse has not been examined. Notably, *Drosophila*, *shot/macf1* mutants, do not have phenotypes reminiscent of loss of *wnt/wingless* mutant phenotypes. The lack of *wnt* phenotypes in *Drosophila*, the large size of the *macf1* gene, and the existence of multiple splice forms of *macf1* present in vertebrates may reflect a novel function for *macf1* in vertebrates or represent a less severe phenotype if some function remains intact. However, these possibilities remain to be investigated.

In oocytes of mammals, the distribution of mitochondria and ER is also dynamic during oogenesis, and mitochondria quality correlates with outcome in fertility clinics (reviewed by Jansen, 2000). It is not clear whether there is an analogous link between the position of the Balbiani body in early oocytes of mammals and the position of the animal–vegetal axis of late-stage oocytes and eggs.

• • • •

Preparing Developmentally Competent Eggs

A developmentally competent egg has the correct number of chromosomes and the right composition of maternal products to sustain its development until expression of the embryo's own genes is activated. This requires proper regulation of nuclear aspects of oocyte development, meiosis, and cytoplasmic aspects, including preparation for activation, fertilization, and early development (e.g., preimplantation in mammals). After germline stem cell division and differentiation, the oocytes initiate meiosis, but then arrest in diplotene of prophase I (Figure 8). During the prolonged prophase I arrest in meiosis, oocytes produce or accumulate mRNAs and proteins necessary to complete later aspects of oogenesis and early development of the embryo (Figure 9). At the same time, the oocyte and the supporting somatic cells undergo dramatic growth and proliferation, respectively. Diplotene arrest is maintained until the organism reaches reproductive maturity, at which point oocytes are periodically induced to resume meiosis during maturation.

NUCLEAR MATURATION: CHROMATIN MODIFICATION

Oocytes that have differentiated enough accumulated nutrients (referred to as post-vitellogenic or oocytes that have accumulated sufficient yolk in nonmammalian vertebrates) are competent to resume meiosis during the process of oocyte maturation. One hallmark of ensuing maturation across phyla is the movement of the oocyte nucleus, also known as the germinal vesicle, to a cortical position. This cortical position is defined as the animal pole of the oocyte. In addition to changes in the position of the nucleus, remodeling occurs within the nucleus as is evident from chromatin morphology. During the metabolically active period of meiotic arrest, the chromatin is in a decondensed conformation, which in mammals (e.g., human, mouse) is known as the nonsurrounded nucleolus stage (Debey et al., 1993). As the oocyte grows and differentiates further, the chromatin becomes more condensed, and in some animals (mouse and humans), the heterochromatin forms a rim around the nucleolus; this is known as the surrounded nucleus conformation (Parfenov et al., 1989; Debey et al., 1993; Zuccotti et al., 1995). Once the oocyte has reached a maturation competent size and

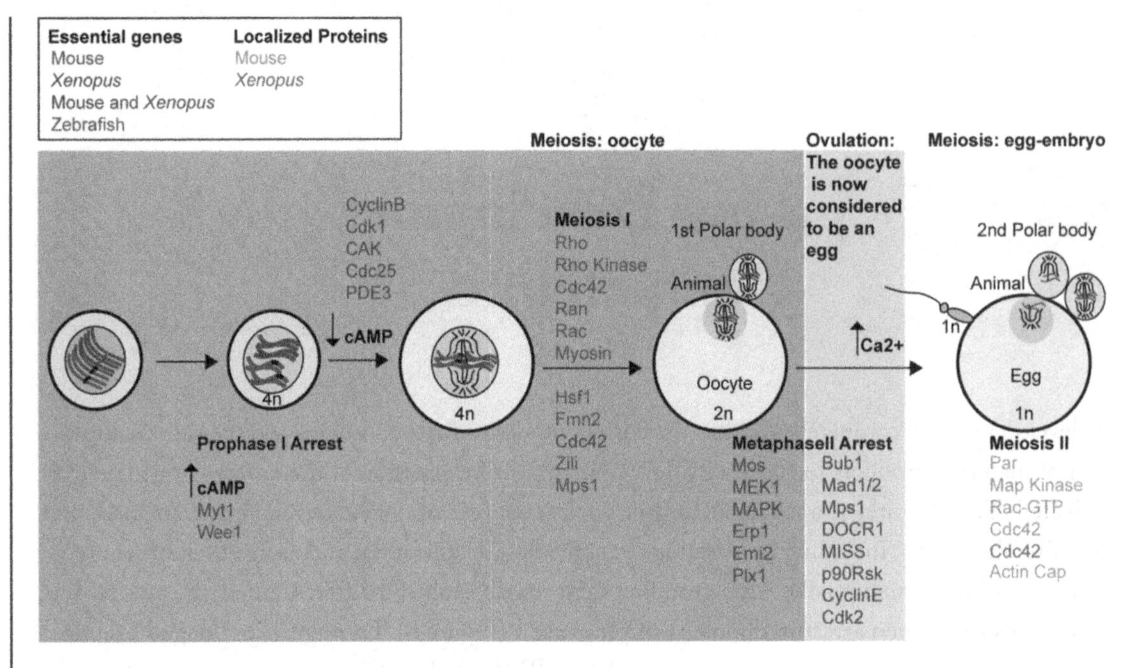

FIGURE 8: Schematic displaying meiosis and the genes whose function is known to be required for normal progression and completion of meiosis to prevent aneuploidy.

molecular composition, it can be induced to resume meiosis. The condensed chromatin conformation accompanies global transcriptional repression and is coincident with meiotic maturation (Parfenov et al., 1989; Schramm et al., 1993; Miyara et al., 2003). However, acquisition of surrounded nucleus conformation is independent of, and is not required for, transcriptional silencing, as demonstrated by the nucleoplasmin knockout mouse, which displays defective nucleolar morphology without disrupting transcriptional repression associated with maturation (De La Fuente et al., 2004). How global chromatin conformational changes and genome silencing in meiotic oocytes are regulated is not understood. It is clear that the timing and duration of the silencing profoundly impacts developmental competence as artificially extending the period of transcriptional silencing compromises developmental potential in the mouse (De La Fuente and Eppig, 2001). Global silencing precedes germinal vesicle break down (GVBD) and chromatin condensation in mouse oocytes; defective silencing in denuded oocytes provided evidence that silencing is mediated by a mechanism involving the cumulus granulosa cells (Colonna, Cecconi et al., 1989; De La Fuente and Eppig, 2001; Eppig, 2001; Matzuk et al., 2002). The signals from the granulosa cells to the follicle cells are not fully understood.

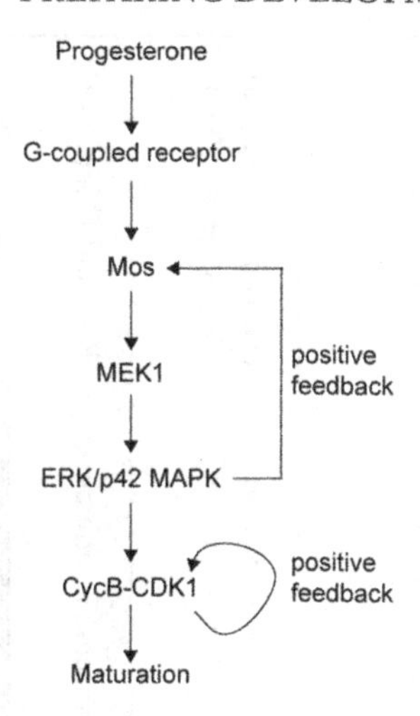

FIGURE 9: Multiple feedback loops ensure that activation of the MPF is all or none.

MATURATION CUES: TO RESUME OR NOT TO RESUME MEIOSIS

In the absence of maturation cues, the default state of oocytes is to remain arrested in meiosis I. When maturation cues are present, repression of competent oocytes is alleviated. The cues that trigger completion of the first meiotic division and elimination of half the DNA in the first polar body are well defined biochemically, and essential regulators of meiosis and maturation have been identified through genetic and interference studies in model systems (Figure 8). Multiple feedback loops ensure that activation is an all-or-none and nonreversible phenomenon. Although the detailed events of meiosis will not be discussed in full, key regulators of early meiosis, maturation, activation, and fertilization of the egg are summarized in Figure 8 and Table 2 (for additional details on the biochemical pathways regulating maturation, the reader is referred to Ferrell, 1999; Ferrell et al., 2009; Masui, 2001). In most animals, gonadotropins induce maturation by triggering activation of the maturation promoting factor (MPF) complex that includes the cell cycle regulators cell division control 2 (Cdc2), also known as cyclin-dependent kinase 1 (Cdk1) and cyclin B (Figure 9) (Labbe et al., 1989; Draetta, 1989; Gautier, 1990; Yamashita, 1992). Before maturation, the MPF complex is maintained as an inactive phosphorylated pre-MPF complex (reviewed by Masui, 2001). Map kinase is responsible for activation of the MPF and like MPF is maintained in an inactive state until maturation. MAPK activation during maturation promotes activity of Cdc25, a phosphatase that promotes activation of MPF and causes nuclear membrane dissolution during the process known as germinal vesicle break down (GVBD) (Haccard et al., 1990; Ferrell et al., 1991).

MAPK regulation of MPF has been implicated in preventing DNA replication between meiosis I and meiosis II, separating the chromosomes during meiotic divisions and blocking parthenogenetic activation (premature activation) during meiosis. The mechanisms by which MAPK

TABLE 2: Maternal regulators of vertebrate development

GENE	MOLECULAR NATURE	PROCESS IMPLICATED	NATURE OF GENE INTERFERENCE	REFERENCES
ANIMAL-VEGETAL POLARITY, MRNA LOCALIZATION				
buckyball	Novel protein	Zebrafish: Balbiani body formation. Animal vegetal oocyte and egg polarity. Germ plasm assembly	ENU Mutant	(Bontems et al., 2009; Dosch et al., 2004; Marlow and Mullins, 2008)
Magellan, macf1	Microtubule actin crosslinking protein	Zebrafish: Balbiani body translocation and disassembly. Oocyte and egg polarity	ENU Mutant	(Dosch et al., 2004; Gupta et al., 2010)
VegT	T-box transcriptional regulator	Xenopus: Vegetal transcript localization in oocytes. Germ layer patterning	Oocyte depletion	(Heasman et al., 2001; Zhang et al., 1998)
Xlsirts	Non-coding RNA	Xenopus: Vegetal transcript localization in oocytes	Antisense interference	(Kloc and Etkin, 1994)
OOCYTE MATURATION, MEIOSIS				
a kinase anchor protein 1 (AKAP1)	A kinase anchor protein		Meiotic maturation	(Newhall et al., 2006)
cell division control protein 42, Cdc42	Small GTPase	Mouse: Oocyte polarity during meiosis Spindle	siRNA	(Cui et al., 2007; Na and Zernicka-Goetz, 2006)
cyclin-dependent kinase 2, cdk2	Cyclin-dependent kinase	Completion of prophase I	Mouse: knoc kout	(Berthet et al., 2003; Ortega et al., 2003)

cell division cycle 25 homolog B, cdc25b	Dual specificity phosphatase	Release from prophase I arrest	Mouse: Mutant	(Lincoln et al., 2002)
cyclic nucleotide phosphodiesterase 3A, pde3a	cAMP hydrolyzing enzyme	Release from prophase I arrest	Mouse: Mutant	(Masciarelli et al., 2004)
c-mos	Serine/threonine kinase	Meiotic arrest/CSF arrest	Mouse: Mutant	(Colledge et al., 1994; Hashimoto et al., 1994)
cytoplasmic polyadenylation element binding protein (cpeb)	cytoplasmic polyad-enylation element binding protein	Mouse: germ cell differentiation, follicle growth	mutant	(Racki and Richter, 2006; Tay and Richter, 2001)
endogenous meiotic inhibitor 2, emi2; erp 1	APC/C inhibitor	Progression to Meiosis II	anti-sense morpholino	(Madgwick et al., 2006; Tay and Richter, 2001)
formin-2	Formin homology gene	Mouse: Spindle polarity and polar body extrusion during Meiosis I, cytokinesis	Mutant	(Dumont et al., 2007; Leader et al., 2002)
Heat shock factor 1 (Hsf1)	Heat shock factor 1	Mouse: nuclear maturation, egg activation fertilization Maturation, cleavages	knockout	(Bierkamp et al., 2010; Christians et al., 2000; Metchat et al., 2009) (Zearfoss et al., 2004)
hermes/rbpms	RNA binding protein		morpholino	
mitotic arrest deficient 2, mad2	spindle checkpoint protein		morpholino	

TABLE 2: (*continued*)

GENE	MOLECULAR NATURE	PROCESS IMPLICATED	NATURE OF GENE INTERFERENCE	REFERENCES
OOCYTE MATURATION, MEIOSIS				
mismatch repair gene MutL homolog 1, Mlh1	Mismatch repair	Mouse: Meiosis Zebrafish: Meiosis	Mouse: mutant Zebrafish: ENU mutant	(Edelmann et al., 1996; Feitsma et al., 2007)
monopolar spindle 1, mps1	kinase	Zebrafish: Meiosis	ENU mutant	(Poss et al., 2004)
over easy	Unknown	Zebrafish: Cytoplasmic maturation	ENU mutant	(Dosch et al., 2004)
rac	Small GTPase	Mouse: Oocyte polarity during meiosis Spindle	siRNA	(Cui et al., 2007)
ruehrei	Unknown	Zebrafish: Cytoplasmic maturation	ENU mutant	(Dosch et al., 2004)
sunny side up	Unknown	Zebrafish: Cytoplasmic maturation	ENU mutant	(Dosch et al., 2004)
souffle	Unknown	Zebrafish: Cytoplasmic maturation	ENU mutant	(Dosch et al., 2004)
t-complex distorter encoding a dynein light chain, tctex/Tex14	t-complex distorter encoding a dynein light chain	Mouse: mutants produce fewer oocytes, ring canal development	knockout	(Caggese et al., 2001)
zili	Piwi-clade of Argonaute (Ago) proteins	Zebrafish: meiosis	ENU mutant	(Houwing et al., 2008)
brom bones/hnRNP1/PTB	Heteronuclear RNA binding protein/poly pyrimidine tract binding protein	Zebrafish: Egg Activation; dorsal axis specification	ENU mutant	(Wagner et al., 2004) (Mei et al., 2009)

EGG ACTIVATION				
Claustro	Unknown	Egg activation/cytoplasmic segregation	ENU mutant	(Pelegri et al., 2004; Pelegri and Schulte Merker, 1999)
emulsion	Unknown	Egg activation/cytoplasmic segregation	ENU mutant	(Dosch et al., 2004)
*FatVg/*adipophilin	Lipid vesicle-associated protein	Cortical rotation, dorsal axis formation, and germ plasm segregation	Oocyte depletion	(Chan et al., 2007)
jumpstart	Unknown	Egg activation	ENU mutant	(Dosch et al., 2004)
tripartite motif 36, trim36	Ubiquitin ligase	*Xenopus*: Cortical rotation	Oocyte depletion	(Cuykendall and Houston 2009)
under repair	Unknown	Egg activation /cytoplasmic segregation	ENU mutant	(Pelegri et al., 2004; Pelegri and Schulte – Merker, 1999)
FERTILIZATION				
basonuclin	Zinc finger protein	Mouse: oocyte morphology. RNA poll an II transcription during oogenesis. Chromatin decondensation and DNA replication at 1 cell stage	Transgenic RNAi oocyte depletion	(Ma et al., 2006)
zona pellucida 1, zp1		Extracellular coat formation	Mouse: mutant	(Hedgepeth et al., 1999)
zona pellucida 2, zp2		Extracellular coat formation and sperm binding	Mouse: mutant	(Carvan et al., 2001)

TABLE 2: (*continued*)

GENE	MOLECULAR NATURE	PROCESS IMPLICATED	NATURE OF GENE INTERFERENCE	REFERENCES
FERTILIZATION				
zona pellucida 3, zp3		Extracellular coat formation and sperm recognition	Mouse: mutant	(Rankin et al., 1996) (Andreuccetti et al., 1999)
phosphatidylinositolglycan class A (pig-a)	GPI-anchor biosynthesis protein	Mouse: sperm/egg fusion	Mouse:knockout mutant	(Alfieri et al., 2003)
folliculogenesis specific basic helix-loop-helix, Fig-a	Transcription factor	Mouse: zona pellucida formation	Mutant	(Soyal et al., 2000)
Ubiquitin Carboxy-terminal Esterase L 1, UCHL 1	Deubiquitinating enzyme	Mouse: blocking polyspermy	Mutant	(Sekiguchi et al., 2006)
MATERNAL REGULATORS OF IMPRINTING				
DNA methyltransferase (cytosine-5) 1o, Dnmt1o	DNA methyltransferase	Maintenance of DNA methylation pattern	Mouse: knockout	(Howell et al., 2001)
DNA methyltransferase(cytosine-5) 1, Dnmt1	DNA methyltransferase	Mouse: maintenance of DNA methylation pattern *Xenopus*: repression of zygotic transcription	Mouse knockout *Xenopus*: de pletion	(Dunican et al., 2008; Stancheva and Meehan, 2000)
DNA methyltransferase 3L, Dnmt3L	DNA methyltransferase	Mouse: establishment of DNA methylation pattern	Mouse: knockout	(Bourc'his et al., 2001)
DNA methyltransferase 3a, Dnmt3a	DNA methyltransferase	Mouse: establishment of DNA methylation pattern	Mouse: knockout	(Kaneda et al., 2004)

Zinc finger protein 57, Zfp57	KRAB zinc finger protein	Mouse: maternal/zygotic establishment and maintenance of specific ICRs	Mouse: knockout Human: mutations	(Li et al., 2008; Mackay et al., 2008)
developmental pluripotency associated 3, Dppa3; Stella; PGC7	DNA/RNA binding protein with a characterized nuclear localization sequence	Mouse: maintenace of paternal specific methylation	Mutant	(Bortvin et al., 2004; Nakamura et al., 2007; Payer et al., 2003)
CLEAVAGE, MITOSIS				
acytokinesis	Unknown	Zebrafish	ENU mutant	(Kishimoto et al., 2004)
atomos	Unknown	Zebrafish: karyokinesis and cytokinesis	ENU mutant	(Dosch et al., 2004)
aura	Unknown	Zebrafish: cytokinesis/membrane	ENU mutant	(Pelegri et al., 2004)
barrette	Unknown	Zebrafish: karyokinesis and cytokinesis/membrane deposition	ENU mutant	(Pelegri et al., 2004)
cellular atoll, Sas6	Centriolar coiled coil protein	Zebrafish: cytokinesis, centrosome assembly	ENU mutant	(Dosch et al., 2004; Yabe et al., 2007)
cellular island	Unknown	Zebrafish: cytokinesis	ENU mutant	(Dosch et al., 2004)
cobblestone	Unknown	Zebrafish: reduced cleavages/fewer cells	ENU mutant	(Pelegri et al., 2004)
Eg5, kinesin-like protein 11	Kinesin motor protein	Mouse: cleavage, spindle	Gene trap insertion	(Castillo and Justice, 2007)
futile cycle	Unknown	Zebrafish: pronuclear congression	ENU mutant	(Dekens et al., 2003; Pelegri et al., 2004)

TABLE 2: (*continued*)

GENE	MOLECULAR NATURE	PROCESS IMPLICATED	NATURE OF GENE INTERFERENCE	REFERENCES
CLEAVAGE, MITOSIS				
geminin	Geminin	Mouse: cytokinesis	Mutant	(Gonzalez et al., 2006; Hara et al., 2006)
indivisible	Unknown	Zebrafish: karyokinesis and cytokinesis	ENU mutant	(Dosch et al., 2004)
irreducible	Unknown	Zebrafish: karyokinesis and cytokinesis	ENU mutant	(Dosch et al., 2004)
mos (map kinase)	kinase	*Xenopus:* M-phase delay first cleavage		(Murakami et al., 1999)
nebel	Unknown	Zebrafish: cytokinesis/ furrow formation	ENU mutant	(Pelegri et al., 2004; Pelegri et al., 1999)
partner of inscrutable, pins	Go Loco G protein regulatory protein	Mouse: spindle	RNAi	(Guo and Gao, 2009)
Post meiotic segregation increased 2, pms2; mulL	Histidine kinase-like ATPase; DNA mismatch repair protein	Mouse: DNA mismatch repair	Mutant	(Gurtu et al., 2002)
RGS14 regulator of G protein signaling 14	GTPase-activating protein	Mouse: mitotic spindle formation	Targeted deletion	(Martin-McCaffrey et al., 2004)
T cell leukemia/ lymphoma 1 gene, TCL 1	Cofactor of Akt1 in cell culture	Cleavage	Mouse: knockout	(Narducci et al., 2002)

weeble	Unknown	Zebrafish: partial cytokinesis	ENU mutant	(Pelegri et al., 2004)
zygotic arrest 1, Zar1	ATP-dependent RNA helicase	Mouse: pronuclear congression, activation of zygotic genes	ENU mutant/ knockout	(Wu et al., 2003)
γ-Tubulin, TUBG1	Tubulin	Mouse: polar body extrusion, karyokinesis, spindle formation	Knockout mouse	(Barrett and Albertini, 2007; Yuba-Kubo et al., 2005)
bo peep	Unknown	Zebrafish: DNA loss/ fragmentation	ENU mutant	(Pelegri et al., 2004)
golden gate	Unknown	Zebrafish: defective chromatin segregation	ENU mutant	(Pelegri et al., 2004)
kwai	Unknown	Zebrafish: defective chromatin segregation	ENU mutant	(Pelegri et al., 2004)
screeching halt	Unknown	Zebrafish: chromatin bridges. Development stalls at MBT (MZT)	ENU mutant	(Wagner et al., 2004)
waldo	Unknown	Zebrafish: DNA loss/fragmentation	ENU mutant	(Pelegri et al., 2004)
MATERNAL TO ZYGOTIC TRANSITION (MZT)/ZYGOTIC GENOME ACTIVATION ZGA/ EMBRYONIC GENOME ACTIVATION (EGA)				
argonaute 2, ago2	*Argonaute*	Mouse: oogenesis, ZGA mRNA decay (subset) Zebrafish: MZ erythrocyte maturation	Knockdown	(Cifuentes et al., 2010; Kaneda et al., 2009; Lykke-Andersen et al., 2008)

TABLE 2: (*continued*)

GENE	MOLECULAR NATURE	PROCESS IMPLICATED	NATURE OF GENE INTERFERENCE	REFERENCES
MATERNAL TO ZYGOTIC TRANSITION (MZT)/ZYGOTIC GENOME ACTIVATION ZGA/ EMBRYONIC GENOME ACTIVATION (EGA)				
autophagy related 5 atg5	Protein binding	Mouse: developmental progression; preimplantation development	Mouse: oocyte knockout	(Tsukamoto et al., 2008a; Tsukamoto et al., 2008b)
brahma related gene, BRG1; SWI/SNF-related, Matrix-associated, Actin-dependent Regulator Chromatin 4, SMARCA4; Sucrose Non Fermentation 2b, SNF2Lb, SNF2b, BAF 190	Helicase	Mouse: maternal zygotic transition-activation of a subset zygotic targets	Null mutant	(Bultman et al., 2000; Bultman et al., 2006)
Bromodomain and WD repeat containing protein, Brwd1; wdr9	Dual bromodomain and WD repeat containing protein	Mouse: meiosis, embryonic genome activation	Mutant	(Huang et al., 2003; Lessard et al., 2004; Philipps et al., 2008)
cyclin A2	Cyclin	Mouse: embryonic genome activation at the one-cell stage	siRNA	(Hara et al., 2005)
CCCTC-binding factor, ctcf	Zinc finger DNA binding protein	Mouse: transcriptional activator, repressor, and epigenetic regulator of imprinting	RNAi	(Wan et al., 2008)
dicer, MiR-430	dsRNA-specific ribonuclease	Zebrafish: maternal zygotic- Deadenylation and decay of maternal transcripts Mouse: meiotic progression/ mRNA clearance at maturation	ENU mutant mouse: ZPC Cre knockout	(Giraldez et al., 2006; Murchison et al., 2007)

dgcr8	Double-stranded RNA binding protein	Mouse: RNA clearance in the embryo	Mutant	(Suh et al., 2010; Wang et al., 2007)
hr6, rad6	Rad-related ubiquitin conjugating DNA repair enzyme	Mouse: development beyond two-cell	Mutant	(Roest et al., 2004)
nucleoplasmin 2 (npm2)	Binds histones	Mouse: exit from the first mitotic division	Mouse: Mutant	(Burns et al., 2003)
Sucrose Non Fermentation 2H, snf2H; SWI/SNF-related, Matrix-associated, Actin-dependent Regulator Chromatin 5, SMARCA5	Helicase	Mouse: EGA activation of a subset of TIF targets	RNAi	(Torres-Padilla and Zernicka-Goetz, 2006)
stem loop binding protein, SLBP	Stem loop binding protein	Mouse: development beyond two-cell stage	Knockout	(Arnold et al., 2008)
TIFα, Transcription Intermediary Factor	Transcription factor	Mouse: EGA	RNAi	(Torres-Padilla and Zernicka-Goetz, 2006)
Zinc finger protein 36 like 2 (TIS11D, ERF2, and BRF2)	Zinc finger protein	Mouse: degradation of maternal mRNAs	Targeted disruption	(Ramos et al., 2004)
EPIBOLY, TISSUE COHESIVENESS, EVL DIFFERENTIATION				
α-Catenin	Armadillo/beta-catenin like repeat protein	*Xenopus:* adhesion in blastula	Oocyte depletion	(Kofron et al., 1997)
bedazzled	Unknown	Zebrafish: epiboly	ENU mutant	(Wagner et al., 2004)

TABLE 2: (*continued*)

GENE	MOLECULAR NATURE	PROCESS IMPLICATED	NATURE OF GENE INTERFERENCE	REFERENCES
EPIBOLY, TISSUE COHESIVENESS, EVL DIFFERENTIATION				
betty boop/Map kap kinase2	Kinase	Zebrafish: epiboly	ENU mutant	(Holloway et al., 2009; Wagner et al., 2004)
E-Cadherin *EP-Cadherin* *E-Cadherin half-baked*	Glycoprotein	Mouse: compaction of ICM *Xenopus:* adhesion during cleavage Zebrafish: epithelial formation and maintenance during epiboly Mouse: adhesion	Mouse: mutant, antibody depletion Zebrafish: ENU mutant	(Larue et al., 1994) (Heasman et al., 1994) (Kane et al., 2005) (De Vries et al., 2004)
EP-CAM, Tacstd	Type I transmembrane glycoprotein	Zebrafish: Epiboly	Zebrafish: retroviral insertion mutant	(Amsterdam and Hopkins, 1999; Slanchev et al., 2009)
Interferon Regulatory Factor 6, Irf6	N-terminal helix-turn helix transcription factor	Zebrafish/*Xenopus*: epiboly	Morpholino	(Sabel, d'Alencon et al., 2009)
janus	Unknown	Zebrafish: blastoderm cohesiveness	Mutant	(Abdelilah and Driever, 1997; Abdelilah et al., 1994)
mission impossible	Unknown	Zebrafish: gastrulation/ epiboly	ENU mutant	(Pelegri et al., 2004)
plakoglobin	Catenin	*Xenopus:* adhesion in blastula	Oocyte depletion	(Kofron et al., 1997)
poky, IKK	IKK kinase	Zebrafish: epiboly, EVL differentiation	ENU mutant	(Fukazawa et al., 2010; Wagner et al., 2004)

slow	Unknown	Zebrafish: epiboly	ENU mutant	(Wagner et al., 2004)
yobo	Unknown	Zebrafish: epiboly	ENU mutant	(Odenthal et al., 1996)
PATTERNING AND MORPHOGENESIS				
Axin	Cytoplasmic regulation of G-protein signaling (RGS) domain	*Xenopus:* dorsal–ventral axis formation	Oocyte depletion	(Kofron et al., 2001)
β-Catenin (Ichabod)	Armadillo/beta-catenin repeat protein	*Xenopus:* dorsal axis specification Zebrafish: dorsal axis specification Mouse: adhesion	Zebrafish: mutant *Xenopus:* oocyte depletion	(Bellipanni et al., 2006; Kelly et al., 1998) (De Vries et al., 2004)
activin-like receptor 8, alk8;lost a fin, laf	TGF-β/BMP receptor	Zebrafish: maternal zygotic ventral specification	ENU mutant	(Mintzer et al., 2001)
blistered	Unknown	Zebrafish: ventral tail vein	ENU mutant	(Wagner et al., 2004)
brom bones/hnRNP1/PTB	Heteronuclear RNA binding protein/ poly pyrimidine tract binding protein	Zebrafish: egg activation; dorsal axis specification	ENU mutant	(Wagner et al., 2004) (Mei et al., 2009)
cAMP response element-binding, CREB	Leucine zipper transcription factor	*Xenopus:* morphogenesis, mesoderm specification	Oocyte depletion	(Sundaram et al., 2003)
dickkopf-related protein 1; dkk 1	Secreted Wnt inhibitor	*Xenopus:* dorsa–ventral axis formation	Oocyte depletion	(Cha, Tadjuidje et al., 2008)
eomesodermin	T box transcription factor	Zebrafish: mesoderm patterning	Morpholino	(Bruce et al., 2003)

TABLE 2: (*continued*)

GENE	MOLECULAR NATURE	PROCESS IMPLICATED	NATURE OF GENE INTERFERENCE	REFERENCES
PATTERNING AND MORPHOGENESIS				
Exostosin-1, Ext1	glycosyl transferase Exostin	*Xenopus:* dorsal axis formation	Oocyte depletion	(Tao, Yokota et al., 2005)
GBP glycogen synthase kinase binding protein	T Cell protooncogene homolog. GSK3 inhibitor	*Xenopus:* dorsal axis specification.	Oocyte depletion	(Yost et al., 1998)
forkhead box protein, foxH1; fastH1, schmalspur (sur)	Forkhead transcription factor	*Xenopus:* dorsal axis formation. Transcriptional repression of nodal signaling (*Xnr5* and *Xnr6*) on the ventral side, and activation of *Xnr3.* Zebrafish: maternal zygotic—mesendoderm formation	Zebrafish: ENU mutant	(Kofron et al., 2004) (Pogoda et al., 2000; Sirotkin et al., 2000)
hecate	Unknown	Zebrafish: dorsal axis specification	ENU mutant	(Lyman Gingerich et al., 2005; Pelegri et al., 2004)
Knypek, kny	Glypican	Zebrafish: gastrulation, neurulation, cilia polarity	MZ mutant	(Borovina, Superina et al., 2010)
lazarus, pbx4	Homeodomain protein	Zebrafish: maternal zygotic—hindbrain formation	ENU mutant	(Waskiewicz et al., 2001)
Ligand of Numb protein-X (Lnx-1)	E3 ubiquitin ligase	Zebrafish: dorsal-ventral axis formation	Morpholino	(Ro and Dawid 2009)

LRP6, low density lipoprotein receptor-related protein 6	Low-density lipoprotein receptor-related protein	*Xenopus:* maintains β-catenin levels in ooctytes by degrading Axin protein. Gastrulation and dorsal axis specification	Oocyte depletion	(Kofron et al., 2007)
ogon, sizzled	Secreted frizzled related inhibitor. BMP antagonist	Zebrafish: maternal zygotic dorsal–vental patterning	ENU mutant	(Miller-Bertoglio et al., 1999; Wagner and Mullins, 2002; Yabe et al., 2003)
one-eyed pinhead (oep)/ EGFCFC	Nodal cofactor	Zebrafish: maternal zygotic. Patterning mesendoderm along the animal-vegetal axis	ENU mutant	(Gritsman et al., 1999)
oval; intraflagellar transport protein 88, Ift88	Tetratrico peptide repeat (TPR)	Zebrafish: cilia formation Hedgehog signaling	MZ mutant	(Huang and Schier 2009)
pol delta 1, flathead, fla	DNA polymerase subunit	Zebrafish: replication, mitotic exit	ENU mutant	(Plaster et al., 2006)
pollywog	Unknown	Zebrafish: rostral and caudal morphogenesis	ENU mutant	(Wagner et al., 2004)
POU domain, class 5, transcription factor 1 pou5f1/ pou2; octamerbinding transcription factor 4oct4;speil ohne grenzen, spg	Homeodomain transcription factor	Zebrafish: maternal zygotic regulation of epiboly endoderm formation Mouse: development prior to blastocyst	Zebrafish: ENU mutant Mouse: morpholino	(Foygel et al., 2008; Lunde et al., 2004; Reim et al., 2004) (Reim and Brand 2006)
pug	Unknown	Zebrafish: posterior body morphogenesis	ENU mutant	(Wagner et al., 2004)
radar;gdf6	TGF-β signaling molecule	Zebrafish: ventral axis specification	ENU mutant	(Goutel et al., 2000; Sidi et al., 2003)

TABLE 2: (*continued*)

GENE	MOLECULAR NATURE	PROCESS IMPLICATED	NATURE OF GENE INTERFERENCE	REFERENCES
PATTERNING AND MORPHOGENESIS				
runt-related transcriptionfactor, runx2bt	Transcription factor	Zebrafish: ventral axis specification	Morpholino	(Flores et al., 2008)
scribble, landlocked	Leucine-rich, PDZ domain protein	Zebrafish: maternal zygotic gastrulation movements	ENU mutant	(Wada et al., 2005)
somitabun, piggy tail, Smad5	Transducer, SMAD	Zebrafish: (*somitabun*) dominant maternal effect ventral fates. (*piggy tail*) maternal zygotic ventral specification	ENU mutant	(Hild et al., 1999; Kramer et al., 2002; Mullins et al., 1996)
SRY-related HMG box transcription factor 3, sox3	SRY-related HMG box transcription factor	*Xenopus:* patterning, restriction of *Xnr5* expression	Antibody interference	*Xenopus:* (Zhang et al., 2004; Zhang et al., 2003)
		Zebrafish: limits expression of the Nodal related *cyclops*		
squint Nodal-related 2	Nodal ligand	Zebrafish: maternal zygotic. Patterning mesendoderm along the animal–vegetal axis	ENU mutant morpholino	(Bennett et al., 2007; Dougan et al., 2003; Feldman et al., 2000; Gore et al., 2005; Pei et al., 2007)
Tokkaebi, tkk; Syntabulin	Kinesin I motor linker protein	Zebrafish: dorsal axis specification	ENU mutant	(Nojima et al., 2010; Nojima et al., 2004)

Trilobite, tri; van gogh	Transmembrane PDZ domain-binding protein	Zebrafish: gastrulation, neurulation, cilia polarity	MZ mutant zygotic mutant plus morpholino	(Borovina, Superina et al., 2010; Jessen et al., 2002)
Vg1; growth differentiation factor 1, Gdf1	TGF-β family ligand	*Xenopus:* gastrulation and axial mesoderm specification	*Xenopus:* oocyte depletion	(Birsoy et al., 2006)
wnt 5; pipetail, ppt	Wnt ligand	*Xenopus:* maintains β-catenin levels in ooctytes by degrading Axin protein.	*Xenopus:* oocyte depletion	*Xenopus:* (Cha et al., 2008)
		Gastrulation and dorsal Zebrafish: gastrulation and neural tube formation	Zebrafish: MZ mutant	Zebrafish: (Ciruna et al.,2006)
wnt11; silberblick, slb	Wnt ligand	*Xenopus:* maintains β-catenin levels in ooctytes by degrading Axin protein.	*Xenopus:* oocyte depletion	*Xenopus:* (Kofron et al., 2007; Tao et al., 2005b)
		Gastrulation and dorsal axis specification Zebrafish: gastrulation and neural tube formation	Zebrafish: MZ mutant	Zebrafish: (Ciruna et al., 2006)
Xfrizzled 7, Xfz7	Wnt receptor	*Xenopus:* dorsal mesoderm, dorsoanterior patterning	Oocyte depletion	(Nasevicius et al., 1998)
Xpace4;subtisilin-kexin like 4	pro-protein convertase	*Xenopus:*mesoderm induction	Oocyte depletion	(Birsoy, Berg et al. 2005)
T- cell factor1, Xtcf1	HMG-box superfamily of DNA-binding proteins	*Xenopus:* dorsal axis. Promotes Wnt targets dorsally and represses Wnt signaling in ventrolateral regions.	Oocyte depletion	(Standley et al., 2006)

TABLE 2: (*continued*)

GENE	MOLECULAR NATURE	PROCESS IMPLICATED	NATURE OF GENE INTERFERENCE	REFERENCES
PATTERNING AND MORPHOGENESIS				
Xtcf3, headless/tcf3, T- cell factor3	HMG-box super-family of DNA-binding proteins	*Xenopus:* represses Wnt signaling in ventral and lateral regions.	*Xenopus:* Oocyte depletion	(Houston et al., 2002)
		Zebrafish: maternal zygotic—head morphogenesis	Zebrafish: ENU mutant	(Kim et al., 2000)
T- cell factor4, Xtcf4	HMG-box super-family of DNA-binding proteins	*Xenopus:* dorsal axis specification. Activation of nodal *Xnr3* and chordin	Oocyte depletion	(Standley et al., 2006)
Zinc finger protein 2, zic2	Odd-paired class transcription factor	*Xenopus: limiting nodal*	Oocyte depletion	(Houston and Wylie, 2005)
SUBCORTICAL MATERNAL COMPLEX AND LINEAGE SPECIFICATION IN THE MOUSE				
C*dx2*	homeobox transcription factor	Mouse: TE specification	Oocyte depletion	(Jedrusik, Parfitt et al. 2008)
filia	Novel	Mouse: spindle	Mutant	(Tong et al., 2000a; Tong et al., 2000b; Zheng and Dean, 2009)
Maternal antigen that embryos require, Mater; Nalp14; nucleotide-binding oligomerization domain, leucine rich repeat and pyrin domain containing 5, Nlrp5	Leucine-rich repeat leucine zipper protein, NACHT	Mouse: reduced expression of EGA genes, progression beyond 2 Cell stage	Null mutant	(Tong et al., 2000a)

Maternal oocyte expressed protein 19, moep19; Factor located in oocytes permitting embryonic development, flopped;ces5	KH-domain family RNA binding protein	Mouse: cytoplasmic lattice formation	Mutant	(Li, Baibakov et al., 2008; Tashiro, Kanai-Azuma et al., 2010)
peptidylarginine deiminase (Padi6)	Peptidyl arginine deiminase	Mouse: cytoplasmic lattice formation	Mutant	(Esposito, Vitale et al., 2007; Yurttas, Vitale et al., 2008)
ADHESION/COHESION				
lysophosphatidic acid 1 (LPA1)	G-protein-coupled-cellsurface receptor	*Xenopus:* cortical actin assembly	Oocyte depletion	(Lloyd et al., 2005)
lysophosphatidic acid 2 (LPA2)	G-protein-coupled-cell surface receptor	*Xenopus:* cortical actin assembly	Oocyte depletion	(Lloyd et al., 2005)
Xflop	G-protein-coupled-cell surface receptor	*Xenopus:* cortical actin assembly	Oocyte depletion	(Tao et al., 2005a; Tao et al., 2007)
GERMLINE SPECIFICATION AND MAINTENANCE				
nanos	RNA binding protein	Zebrafish: PGC maintenance	ENU mutant	(Draper et al., 2007)
PERSISTING MATERNAL FUNCTION				
aldehyde dehydrogenase 2, aldh2	Retinaldehyde dehydrogenase	Zebrafish: endoderm development	Mutant plus inhibitor	(Alexa, Choe et al. 2009)
flat head/pol delta 1	*pol delta 1*	Zebrafish: viability	Mutant / ENU, insertion	(Plaster et al., 2006)

TABLE 2: (*continued*)

GENE	MOLECULAR NATURE	PROCESS IMPLICATED	NATURE OF GENE INTERFERENCE	REFERENCES
PERSISTING MATERNAL FUNCTION				
phosphoribosylglycinamide formyltransferase, phosphoribosylglycinamide synthetase, gart	Phosphoribosylglycinamide formyltransferase, phosphoribosylglycinamide synthetase	Zebrafish: maternal zygtoic: pigmentation, axis	Mutant	(Ng, Uribe et al., 2009)
mcm5, minichromosome maintenance 5	AAA ATPase	Zebrafish: replication, phase progression/ mitotic exit	SENU mutant	(Ryu et al., 2005)
misty somites	Novel	Zebrafish: somite boundary formation	Insertion	(Kotani and Kawakami, 2008)
phosphoribosylaminoimidazole carboxylase, phosphoribosylaminoimidazole succinocarboxamide synthetase, paics	Phosphoribosylaminoimidazole Succinocarboxamide synthetase	Zebrafish: maternalzygtoic: pigmentation	Mutant	(Ng, Uribe et al., 2009)
topoisomerase 2A	*topoisomerase*	Zebrafish: viability	Mutant insertion	(Dovey et al., 2009)

activates cyclin kinase activity are diverse and include repression of inactivating kinases, regulation of the cellular distribution of proteins, and regulation of translation (see Abrieu et al., 2001; Masui, 2001). However, the extent to which converting an inactivate pre-MPF to an active complex, *de novo* translation of *cyclins*, or cyclin-regulated proteins, such as Cdc2, contribute to or determine MAPK's role in each of these distinct activities has not been fully elucidated. Nuclear maturation completes with the primary meiotic division and cytokinesis to extrude the first polar body and is followed by arrest of the oocyte in metaphase of meiosis II.

The mature oocyte or egg (if it is ovulated) will remain arrested in meiosis II until it is fertilized and activated or degenerates. Unlike the long-term arrest of immature oocytes, which can be sustained for many months, arrest of mature oocytes is more short term and is maintained for hours. Whereas MPF promotes maturation and resumption of meiosis, cytostatic factor (CSF) promotes M-phase arrest of oocytes and of early embryos (Masui and Markert, 1971) and reviewed by Masui (2001). Failure of oocytes to arrest in meiosis II when the serine/threonine kinase that activates MAP kinase, Mos, is disrupted and mitotic arrest caused by ectopic Mos in *Xenopus* embryos demonstrated the role for CSF in promoting M-phase arrest (Colledge, 1994; Hashimoto, 1994). The consequence to the embryo of disrupting either of these regulatory events is failure to properly segregate the chromosomes in meiosis. Defective meiosis causes aneuploidy and embryonic arrest due to developmental abnormalities. Thus, preventing meiosis from proceeding through to completion before activation of the egg is as important for developmental competence as initiating its resumption.

POLARITY PROTEINS REGULATE THE CYTOSKELETON AND MEIOSIS

In the previous section, the signaling pathways that induce meiosis and maintain arrest were discussed. As mentioned, the nucleus moves from the center of the oocyte to the cortex before the nuclear membrane breaks down; thus, positions the chromatin asymmetrically near the cortex before the meiotic division occurs. Meiosis is an inherently polarized process in all animals whereby asymmetric cell divisions produce daughter cells of unequal size, a huge oocyte cell and two comparatively tiny polar bodies (Figure 8). Underlying this asymmetric cell division is a polarized distribution of the acentriolar meiotic spindle and chromatin. In *Xenopus*, Cdc42 is activated at the cortex above the spindle pole, and interference studies support its essential role in formation of the first polar body and cytokinesis (Figure 10) (Ma et al., 2006a). Similarly, dysregulation of actin myosin contractility by blocking the small GTPase, Rho, or its target, Rho kinase, disrupts germinal vesicle breakdown, spindle rotation, and emission of the first polar body (Zhong et al., 2005). Exposing oocytes to Rho inhibitor at later studies revealed additional roles for Rho in promoting extrusion of the second polar body. In these studies, failed polar body elimination resulted in binucleate zygotes, demonstrating that although Rho regulates several actin-mediated processes in oocytes, it does not

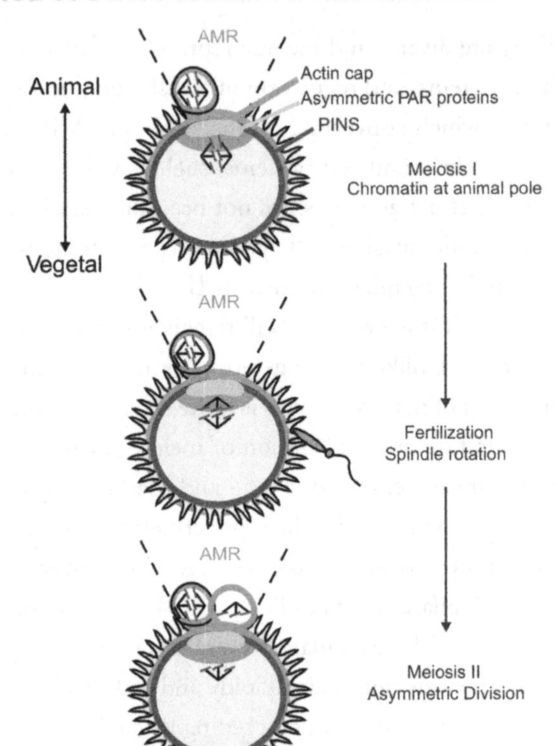

FIGURE 10: Schematic of spindle orientation and asymmetric meiotic divisions of mouse oocytes and eggs. Proteins including PARs localize in the vicinity of the actin cap, while partner of inscrutible PINS is excluded from the cap region. These asymmetrically positioned proteins play essential roles in orienting the spindle and polar body elimination to produce a large oocyte and small polar bodies during the asymmetric meiotic divisions.

appear to directly regulate chromosome separation during meiosis (Moore, Ayabe et al., 1994; Zhong et al., 2005).

In the mouse, an actin cap is organized above the chromatin positioned at the animal pole. Key regulators of cell polarity in multiple cell types (e.g., Par proteins, Map kinase, Rac-GTP, which is the the active form of Rac) also accumulate asymmetrically at this region of the cell cortex (Figure 10) (Deng et al., 2007; Deng et al., 2005; Duncan et al., 2005; Vinot et al., 2004). In the eggs of mice, ectopic actin caps can be induced by inhibiting microtubules with nocodozole, which results in scattering of the chromosomes or adding excess DNA to eggs (Deng et al., 2007; Deng et al., 2005; Halet and Carroll, 2007). The small GTPases, Cdc42, Rac, and Ran, are known to regulate the cytoskeleton in other contexts, and their enrichment in the amicrovillar animal cap region hinted at their involvement in regulating asymmetric protein distribution and spindle polarization. Interference with these GTPases provided functional support for their essential contributions to actin cap formation, spindle polarization, and polar body extrusion during meiosis (Cui et al., 2007; Deng et al., 2007; Halet and Carroll, 2007; Na and Zernicka-Goetz, 2006). Evidence that Rac participates in polar body emission by modifying the existing pool of actin possibly acting through other proteins, such as myosin comes from studies using dominant negative Rac, which disrupts meiosis without impairing actin cap formation (Halet and Carroll, 2007).

Formin2, *fmn2*, nucleates unbranched actin filaments and is required for extrusion of the first polar body and female fertility in the mouse (Leader et al., 2002). Despite its role in meiosis, an apparent lack of a meiotic checkpoint permits *fmn2* mutant females to produce polyploid progeny, most of which arrest at various midgestation stages (Leader et al., 2002). Initial analysis of the mechanism responsible for failed polar body elimination in *fmn2* mutant oocytes indicated defective actin-dependent chromosome movement to the cortex of mutant oocytes when treated with nocodozole (Leader et al., 2002). However, follow up analyses indicated that failure to extrude the polar body was a consequence of impaired cytokinesis during the meiotic division; a defect associated with abnormal localization of Phosphorylated-Myosin II between the chromosomes (i.e., the active form of Myosin) (Dumont et al., 2007). The finding that Fmn2 regulates polar body elimination independent of chromosome segregation is reminiscent of Rho kinase, a known activator of myosin discussed above.

In contrast to the actin cap localization of the Par proteins and GTPases discussed above, partner of inscrutable, Pins, a conserved mediator of polarity and spindle organization (Colombo, Grill et al., 2003; Betschinger and Knoblich, 2004), has the opposite localization in maturation stage oocytes of mice. Pins localizes in the regions that contain the cortical granules and microvilli (Guo and Gao, 2009). Gain of function and RNAi interference with *pins* in mouse oocytes illustrated its role in regulating spindle morphology, chromosome congression and alignment, and association of Tubulin with the acentriolar spindle, during meiosis I (Guo and Gao, 2009). Rescue studies with truncated proteins provided evidence that the asymmetric localization of Pins protein is mediated by its carboxy-terminus, whereas the amino-terminal half of the protein regulates spindle assembly (Guo and Gao, 2009). The mechanisms mediating the opposite localization of Pins and the animal pole positioned polarity proteins in mouse oocytes, and whether complexes present in the amicrovillar animal pole exclude Pins from the actin cap region are not known. Together these inhibitor and mutant studies highlight the conserved mechanisms involving small GTPases and modulators of the cytoskeleton that are employed to accomplish meiotic division and prevent aneuploidy in oocytes of diverse species.

HEAT SHOCK FACTOR 1: PROMOTING NUCLEAR MATURATION AND DEVELOPMENTAL COMPETENCE

Heat shock factor 1 is a conserved transcriptional regulator of the heat shock response pathway. In the oocytes of mice Hsf1 regulates the expression of several chaperones and controls meiotic maturation of oocytes (Metchat et al., 2009; Christians et al., 2000). *hsf1* is dispensable for oocyte development prior to maturation stages; however, during post-ovulation stages *hsf1* mediates egg activation and fertilization (Bierkamp et al., 2010). Evidence that Hsf1 regulates maturation includes impaired germinal vesicle breakdown and meiosis l due to compromised Cdk1 and MapK activity and diminished heat shock protein expression in mutant oocytes (Metchat et al., 2009). In a follow-up study to further explore the basis of the post–ovulation *hsf1* maternal-effect lethality phenotype, Bierkamp and colleagues observed an increased incidence of parthenogenesis, or spontaneous activation of the egg before fertilization, which results in impaired developmental competence and apoptosis at the two-cell stage, as indicated by elevated Caspase 3 (Bierkamp et al., 2010). Spontaneous activation of *hsf1* mutant eggs does not prevent fertilization or polyspermy likely because *hsf1* is also necessary for mediating cortical granule exocytosis during egg activation (Bierkamp et al. 2010).

Further indications of impaired developmental competence in *hsf* mutants include aberrant mitochondria morphology, and likely function, associated with increased reactive oxidative species (ROS) and oxidative stress (Bierkamp et al., 2010). It is unclear how *hsf1* mutant oocytes escape earlier arrest. One possibility is that mutant oocytes may be refractory to elevated ROS and oxidative stress, whereas in the ovary, as indicators of cell death and atresia (i.e., follicle degeneration prior to reaching maturity) were not detected in preovulatory stages (Bierkamp et al., 2010). In this scenario, factors provided by the somatic cells of the follicle may protect the oocyte until ovulation. Alternatively, the damage accumulated prior to ovulation may be below the threshold required to elicit programmed cell death in oocytes, or checkpoints only begin to function in post-ovulatory oocytes when *hsf1* mutant oocytes arrest and undergo apoptosis. Distinguishing between protective mechanisms in the follicle or whether *hsf1* mutant oocytes are simply unable to tolerate the accumulation of ROS and damage to the mitochondria will require further investigation.

Piwis: SILENCING TO PREVENT ANEUPLOIDY

The Piwi-clade of Argonaute (Ago) proteins and their associated piRNAs are regulators of gene and transposon silencing (to protect the genome from damage) in the germline of invertebrates and vertebrates (Cox et al., 1998; Cox et al., 2000; Kuramochi-Miyagawa et al., 2001; Szakmary et al., 2005; Lau et al., 2006; Megosh et al., 2006; Brennecke et al., 2007; Houwing et al., 2007; Seto et al., 2007; Houwing et al., 2008; Kuramochi-Miyagawa et al., 2008; Lau et al., 2009). PiRNAs are 24- to 31-bp RNAs produced by a distinct mechanism from MiRNAs (discussed later in the chapter on silencing maternal messages). Instead of being produced from dsRNA, piRNAs are processed from primary transcripts that are derived from active transposable elements or repetitive chromosomal loci known as piRNA clusters (Aravin et al., 2007; Brennecke et al., 2007). The primary piRNAs exist in two forms (Ago associated and Aubergine/PIWI associated). Cleavage by Slicer inactivates the piRNA target and, at the same time, generates a new piRNA (i.e., piRNAs are generated via a feed-forward or ping-pong amplification cycle) (Brennecke et al., 2007; Carmell et al., 2007; Kato et al., 2007; Kuramochi-Miyagawa et al., 2008).

The consequences of disrupting Piwi proteins vary among species; mutations disrupting Piwi have been correlated with elevated transposon activity, with compromised histone methylation, diminished *de novo* DNA methylation, impaired control of translation (Kuramochi-Miyagawa et al., 2008; Aravin and Bourc'his, 2008; Carmell et al., 2007; Deng and Lin, 2002; Grivna et al., 2006; Houwing et al., 2008; Houwing et al., 2007; Kuramochi-Miyagawa et al., 2004). In the mouse, Miwi, Miwi2, and Mili are essential in the male germline for meiosis and differentiation, although only Miwi2 maintains germline stem cells in the testis (Carmell et al., 2007; Deng and Lin, 2002; Kuramochi-Miyagawa et al., 2004). Zebrafish mutants revealed earlier roles for Piwi proteins in maintaining premeiotic germ cells between larval stages and adulthood. In these mutants, the germline stem cells fail to enter meiosis and are consequently lost to apoptosis (Houwing et al., 2008; Houwing et al., 2007). Mutants with null alleles develop exclusively as sterile males because the germline is required for ovary development in zebrafish (Houwing et al., 2008; Houwing et al., 2007; Weidinger et al., 2003). Maternal-effect egg phenotypes of hypomorphic *zili* alleles indicate that Zili is also required for female meiosis as is evident from failed polar body elimination from mutant oocytes (Houwing et al., 2008). The opposite sexually dimorphic requirement for Piwi proteins in regulating female meiosis in fish and male meiosis in mouse is intriguing. The meiosis defects are associated with increased transposon activity in mouse, but seem to be transposon-independent in zebrafish based on comparable transcript abundance in mutant and wild-type ovaries (Carmell et al., 2007; Deng and Lin, 2002; Houwing et al., 2008; Houwing et al., 2007; Kuramochi-Miyagawa et al., 2004). The transposon-independent mechanism regulating meiosis in female zebrafish, and whether transposon-independent activity of Piwi also contributes to regulation of meiosis in male mice remain to be determined.

mlh1/mps1: MISMATCH REPAIR AND MEIOSIS

The mismatch repair gene mutL homolog 1, *mlh1*, also known as colon cancer, nonpolyposis type 2, regulates meiosis and developmental competence in zebrafish and in the mouse as is evident from the maternal-effect morphological defects and arrest phenotypes caused by defective chromosome separation during meiosis (Edelmann et al., 1996; Feitsma et al., 2007). The progeny of mutant females show variable numbers of maternal chromosome copy (zero to two copies; (Feitsma et al., 2007). Remarkably, zebrafish embryos with two maternal chromosomes, or those that are triploid, progress further through development compared with their siblings without maternal chromosomes (Feitsma, Leal et al., 2007). The finding that in zebrafish an excess of chromosomes is tolerated much better than a deficiency of maternal chromosomes is evidence that checkpoints are not operational prior to activation of the zygotic genome. The earlier activation of the zygotic genome and presumably checkpoints at the two-cell stage in mouse likely accounts for the comparatively advanced developmental stage attained by zebrafish progeny of mutant mothers.

In zebrafish *monopolar spindle1*, *mps1*, kinase is abundant in premeiotic germ cells and early meiotic stage, but not late-stage oocytes and sperm (Poss et al., 2004). Mutants with a temperature-sensitive allele of *mps1* are viable to adulthood at permissive temperatures (Poss, 2004; Poss et al., 2004). The progeny of a mutant male or a mutant female arrest in development is due to aneuploidy; thus, Mps1 is essential to prevent meiotic nondisjunction during meiosis in the germline (Poss, 2004; Poss et al., 2004). The expression of *mps1* in early germ cells and the presence of centrosomes and bipolar spindle in the embryos derived from *mps1* mutant parents indicate that Mps1 is required in the germline during meiosis to later promote normal development of the embryo (Poss et al., 2004). Thus, the paternal and maternal-effect phenotypes of *mps1* implicate a mitotic checkpoint in ensuring euploidy in zebrafish.

CYTOPLASMIC MATURATION: CLEARING THE YOLK IN FISH

In addition to nuclear maturation, cytoplasmic changes contribute to production of a viable egg (Figure 9) (reviewed by Eppig, 1996). In zebrafish oocytes, the clarity of the yolk is a cytoplasmic indicator of oocyte maturation (Wallace and Selman, 1990). Prior to maturation, the oocyte is opaque due to the accumulation of membrane-enclosed crystalline yolk (Figure 11) (Selman et al., 1993). At the end of maturation the yolk becomes noncrystalline and translucent (Figure 9) (Wallace and Selman, 1990). The change in the yolk appearance is attributed to, and correlates with, cleavage and processing of the major yolk proteins. Several mutants disrupting yolk maturation have been identified through maternal-effect screens in the zebrafish (Dosch et al., 2004). The eggs from these mutant mothers are opaque and the animal pole positioned blastodisc is not readily apparent; thus, mutant eggs from *over easy*, *ruehrei*, *soufflé*, and *sunny side up* mutant mothers resemble opaque oocytes rather than the translucent eggs of wild type with a conspicuous blastodisc at the animal pole (Figure 9) (Dosch et al., 2004). Whether the disrupted genes specifically function in yolk maturation or also contribute to meiosis or nuclear maturation has not been determined. The molecular identity of the disrupted genes in these mutants will reveal whether they regulate distinct steps of a single pathway dedicated to regulating yolk maturation or are components of a pathway more broadly required for oocyte maturation.

After nuclear and cytoplasmic maturation, oocytes arrests in meiosis II. As discussed earlier, in many animals, including the model systems discussed here, this second meiotic arrest is

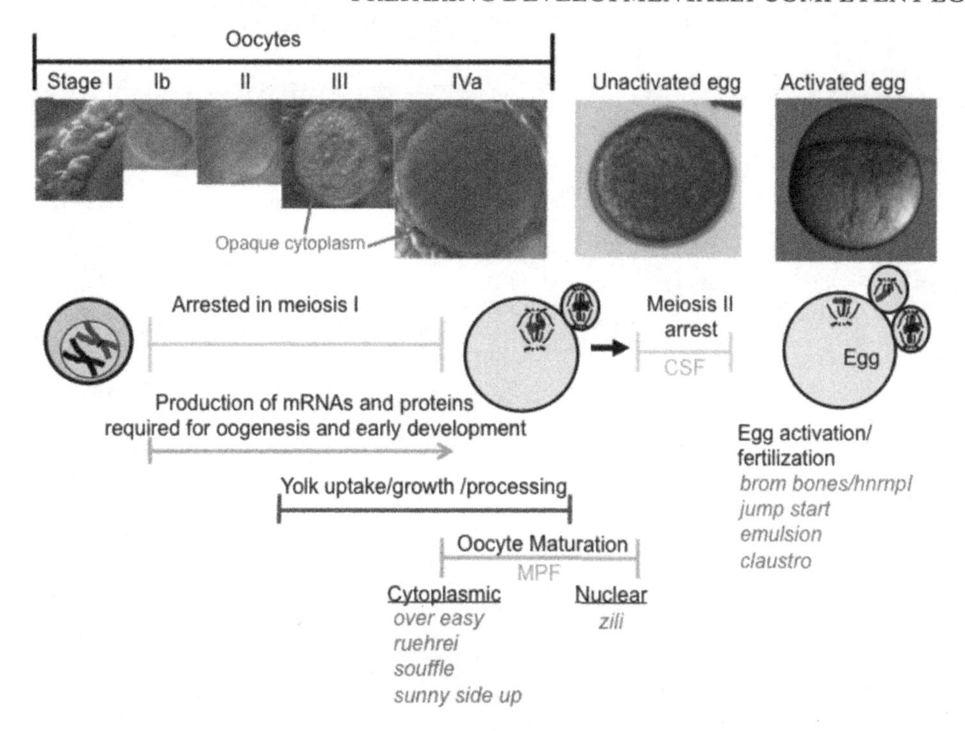

FIGURE 11: Oocyte development in zebrafish and the genes identified through forward genetic screens that regulate distinct aspects of maturation and egg activation to generate a developmentally competent egg. Prior to maturation the cytoplasm of zebrafish eggs is opaque. After maturation, the yolk is translucent due to cleavage of the major yolk proteins. Maturation promoting factor (MPF) is activated to alleviate meiotic arrest. After the first polar body is eliminated during meiosis I cytostatic factor (CSF) maintains meiosis II arrest until the egg is activated. Oocyte images are whole mount views acquired with a compound microscope. Images of eggs are lateral views, animal pole to the top acquired with a dissecting microscope.

maintained during and even after ovulation. Ovulation is believed to involve epigenetic changes and activities independent of macrophage migration inhibitory factor, MIF, because exogenously supplied MIF is sufficient to induce maturation but not ovulation in fish and frogs (reviewed by De La Fuente, 2006; Clelland and Peng, 2009). Ovulation commences when the follicle cell layer basement membranes degrade and release the oocyte into the oviduct. Once ovulated, the oocyte is considered to be an egg.

• • • •

Egg Activation

In many animals (e.g., flies, worms, frogs, mice), fertilization triggers activation of the egg, whereas in others, including zebrafish, contact of the egg with spawning media (e.g., water) stimulates the egg to activate development. Egg activation is initiated by or temporally coincident with sperm entry in many animals, therefore, the fertilized egg must first complete the second meiotic division to form the haploid maternal pronucleus then mediate fusion of the male and female pronucleus while transitioning from a meiotic cell cycle to the mitotic cell cycle of the now diploid embryo.

Egg activation, whether initiated by or independent of fertilization, induces a rise in intracellular calcium. This calcium increase often manifests as a signaling wave that traverses the egg and releases it from meiotic arrest and metabolic dormancy. Once activated, the egg will resume and complete meiosis II with the release of the second polar body (Figure 11). In addition to restarting meiosis, the calcium signal mediates dramatic rearrangements of the egg surface, including the cortical reaction to block polyspermy and the rearrangement of cytoplasm prior to the first cleavage.

The cortical reaction prevents polyspermy during fertilization in animals including vertebrates (reviewed by Gardner and Evans, 2006; Tsaadon et al., 2006; Wong and Wessel, 2006). This shared aspect of egg activation involves calcium-triggered exocytosis of cortical granules (CGs), large secretory vesicles, docked at the oocyte cortex during oogenesis (Figure 12) (reviewed by Tsaadon et al., 2006; Wong and Wessel, 2006). The CGs contain enzymes and structural proteins that are released into the perivitelline space to promote remodeling of the vitelline envelope surface, driving elevation and hardening of the vitelline envelope, referred to as the chorion in zebrafish and zona pellucida in mammals (Figure 13). The hardened chorion and the space created as the chorion elevates or lifts away from the surface of the egg provide a protective barrier for the egg (reviewed by Hart, 1990). The space between the chorion and the vitelline envelope is filled with water from the environment and the contents of the exocytosed cortical granules, the so-called perivitelline fluid which serves as a buffer between the egg and the external environment.

Although rearrangement of cytoplasm during egg activation is also a common theme in eggs, the mechanisms required to accomplish cytoplasmic rearrangement can be quite distinct in different animals. In *Xenopus*, a dorsal-ward shift of the cortex, known as cortical rotation, is mediated by

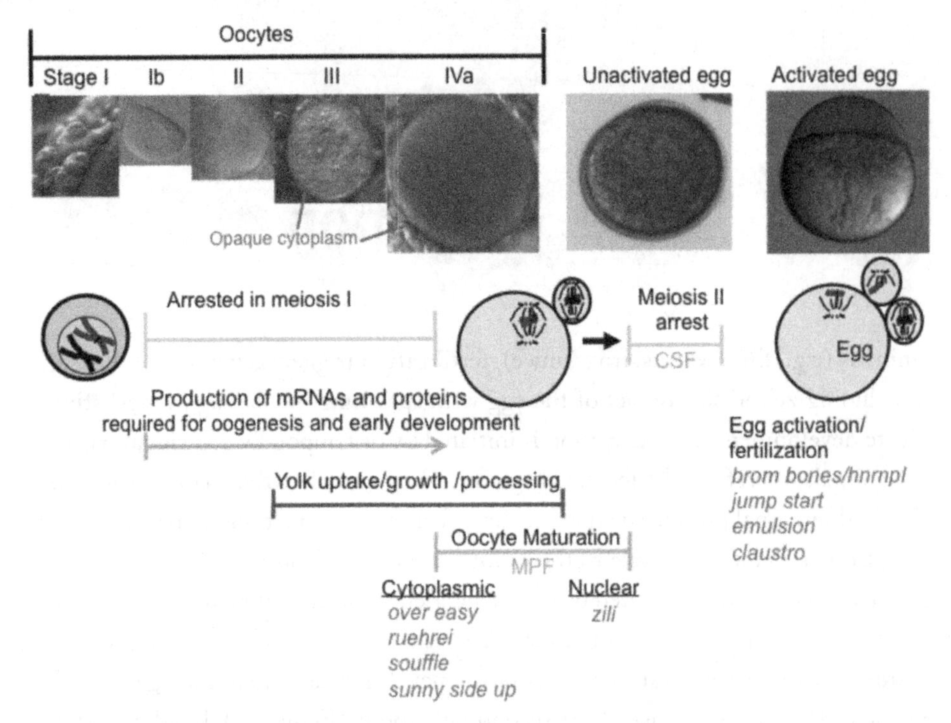

FIGURE 11: Oocyte development in zebrafish and the genes identified through forward genetic screens that regulate distinct aspects of maturation and egg activation to generate a developmentally competent egg. Prior to maturation the cytoplasm of zebrafish eggs is opaque. After maturation, the yolk is translucent due to cleavage of the major yolk proteins. Maturation promoting factor (MPF) is activated to alleviate meiotic arrest. After the first polar body is eliminated during meiosis I cytostatic factor (CSF) maintains meiosis II arrest until the egg is activated. Oocyte images are whole mount views acquired with a compound microscope. Images of eggs are lateral views, animal pole to the top acquired with a dissecting microscope.

microtubules and plays a key role in organizing the cytoplasm of the embryo (Figure 14). In zebrafish, the yolk and cytoplasm are largely intermingled prior to egg activation. actin- and myosin-associated streaming of the cytoplasm, also known as ooplasm, through "channels" between the yolk globules, is thought to mediate partitioning of the cytoplasm and formation of the blastodisc at the animal pole (Figure 14) (Becker and Hart, 1999; Fernandez et al., 2006; Hart and Fluck, 1995; Leung et al., 2000). In zebrafish, movement of the cytoplasm is bipolar, occurring both toward the

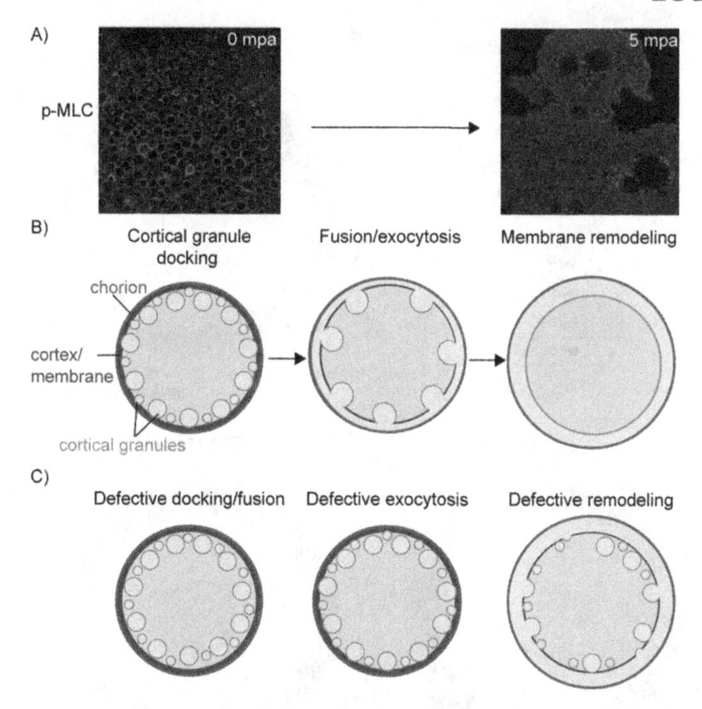

FIGURE 12: Cortical granule exocytosis contributes to egg activation. (A) Phosphorylated (active)-myosin regulatory light chain marks the cortical granules prior to egg activation and the remodeled cell surface following completion of cortical granule exocytosis in zebrafish eggs. (B) Schematic cross section depicting the events of cortical granule exocytosis. The cortical granules move to and dock at the cortex in late-stage oocytes. Upon egg activation, the cortical granules fuse with the egg membrane and release their contents into the perivitelline space, driving chorion elevation and remodeling of the cell surface. (C) Schematic cross section views depicting cortical granule related egg activation defects observed in zebrafish maternal-effect mutants. Defective translocation or docking of the cortical granules can compromise their ability to fuse with the cortex upon activation. Failure of the cortical granules to fuse and exocytose their content or reduced exocytosis results in persisting cortical granules. Both defective docking and defective exocytosis would cause diminished chorion elevation. Finally, cortical granule exocytosis and chorion elevation can be normal while remodeling of the cortex is compromised. Mpa = minutes post-activation.

animal and the vegetal poles, but animal pole axial streaming is faster (Leung et al., 2000). Maternal products protect the egg from being fertilized by excess sperm and from undergoing premature or incomplete activation (Webb et al., 1995). The following subchapters highlight maternal-effect mutants that compromise events during the transition from egg to embryo thus limit developmental potential of the embryo.

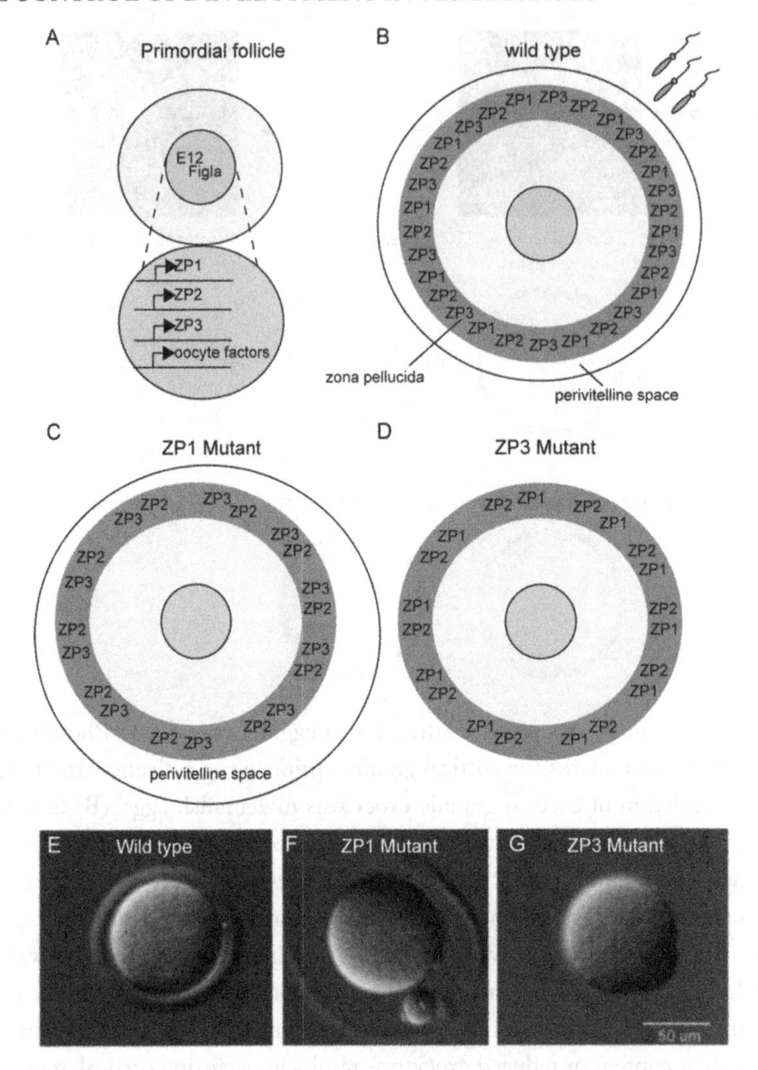

FIGURE 13: (A) Figla promotes expression of the three mammalian zona pellucida proteins. (B–D) Each uniquely contribute to the formation, function, and integrity of the zona as can be seen (F, G) in the mouse mutant phenotypes. (E–G) Images of unfixed ovulated eggs from Rankin et al. (2000).

ZEBRAFISH EGG ACTIVATION MUTANTS

Cortical granule exocytosis, elevation of the chorion, and blastodisc elevation during cytoplasmic segregation are hallmarks of egg activation in zebrafish. Evidence that hnRNPI promotes calcium induction necessary for egg activation in zebrafish comes from compromised egg activation phenotypes of *brom bones* (*hnRNPI*) maternal-effect mutants associated with compromised calcium

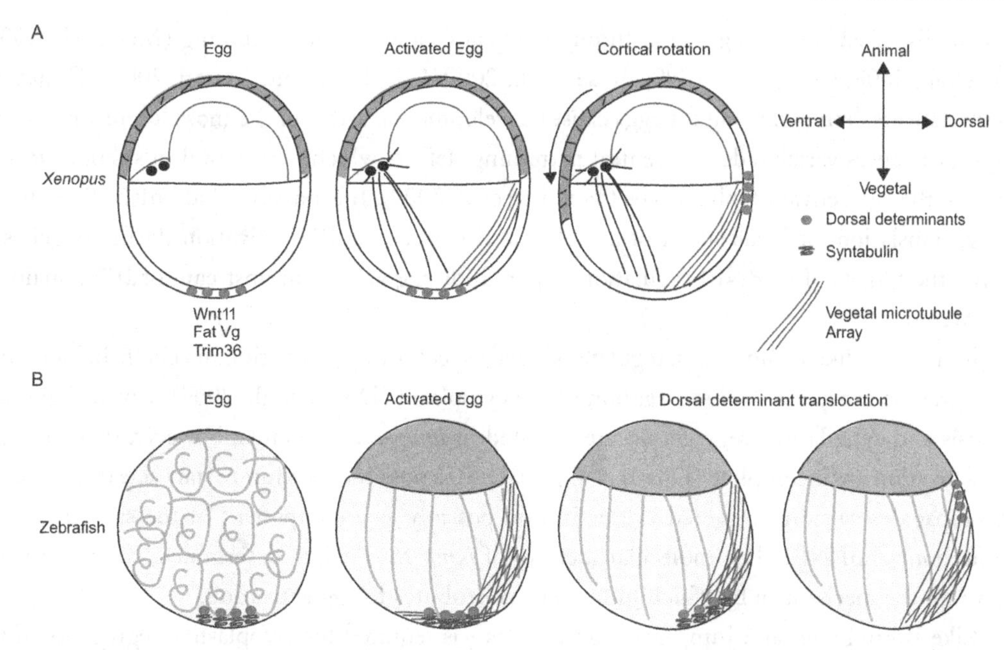

FIGURE 14: Mechanisms to move the cytoplasm and dorsal determinants in (A) *Xenopus* and (B) zebrafish. (A) The dorsal determinants are positioned at the vegetal pole of *Xenopus* oocytes. Wnt11 and Fat Vg are also positioned at the vegetal pole along with Trim36. Wnt11 induces dorsal determination, while Fat Vg contributes to trafficking of the dorsal determinants toward the dorsal side during cortical rotation. Trim36 is required for assembly of the vegetal microtubule array (red), which drives cortical rotation relative to the site of sperm entry. Cortical rotation contributes to positioning the dorsal determinants on the prospective dorsal side. (B) In zebrafish, Syntabulin is positioned at the vegetal cortex of oocytes and eggs, and is presumably tethering the dorsal determinants in the vegetal pole. Upon activation, a vegetal microtubule array (red) is assembled. Syntabulin localization shifts asymmetrically, but remains in the vegetal cortex. Later, Syntabulin protein degradation is thought to liberate the dorsal determinants to travel along the microtubule array toward the blastoderm where they will specify dorsal. Incomplete egg activation in *brom bones* mutants blocks dorsal determinant translocation. Disrupting *hecate* also blocks dorsal determinant translocation by an unknown mechanism. Schematics are lateral, cross-section views with prospective dorsal to the right.

as visualized by the bioluminescent calcium-sensitive protein Aququeporin (Mei et al. 2009). Rescue of the maternal-effect egg activation defects by injecting either calcium or IP3 provided additional evidence that Brom bones promotes inositol (1,4,5-triphosphate) signaling and, consequently, calcium induction necessary for optimal egg activation (Mei et al., 2009). In *Drosophila* and *Xenopus* hnRNPI regulates translation and localization of mRNAs in oocytes; however, in zebrafish, Brom bones are dispensable for progression through oogenesis and oocyte patterning (Mei et al., 2009; Lewis et al., 2008; Cote et al., 1999; Besse et al., 2009; Czaplinski and Mattaj, 2006). Failure to properly activate *brom bones* mutant eggs causes developmental arrest in the most severely impacted embryos, or causes variable dorsal–ventral patterning defects which correspond in strength to the severity of the egg activation phenotype (Mei, Lee et al. 2009). It is possible that hnRNPI regulates splicing, translation, or localization of a molecule necessary for IP3 activation during oogenesis; however, the targets of hnRNPI regulation required for egg activation upstream of IP3 remain to be discovered.

Jump-start, like Brom bones, regulates several aspects of egg activation in zebrafish, including cortical granule exocytosis and segregation of the cytoplasm (Dosch et al., 2004). Cortical granule exocytosis and cytoplasmic segregation are initiated in *jump-start* mutants; however, these aspects of egg activation are incomplete (Dosch et al., 2004). It is possible that Jump-start acts downstream of or propagates calcium-triggered egg activation possibly as a component of the same pathway as Brom bones/hnRNPI. The molecular identity of *jump-start*, when determined, should provide insight into the mechanism by which Jump-start contributes to egg activation.

Like Brom bones and Jump-start, under repair is required for cytoplasmic segregation during egg activation (Pelegri et al., 2004). Normal chorion elevation in *under repair* mutant progeny suggests it might act specifically in cytoplasmic segregation to the blastodisc rather than regulation of egg activation in general (Pelegri et al., 2004). Actin and myosin and their regulators have been implicated in cytoplasmic segregation through inhibitor studies. Inhibiting actin assembly with cytochalasin B impairs constriction at the blastoderm margin in the animal hemisphere, diminishes animal–vegetal cytoplasmic streamers, slows the flow of tracer microspheres, and impedes blastodisc formation (Leung et al., 1998; Leung et al., 2000). Inhibiting microtubules with colchicine does not block cytoplasmic segregation during early egg activation, but instead disrupts later cleavages and causes embryonic arrest (Leung et al., 1998; Leung et al., 2000).

In Medaka, cytochalasin B and colchicine treatment both diminish cytoplasmic segregation, but in this case, microtubules seem to be more involved in directionality of the streamers. It is possible that *under repair* disrupts upstream regulators or downstream targets of actin/myosin contractility during egg activation. However, distinguishing whether *under repair* is necessary for actin/myosin signaling or, like *brom bones*, impairs calcium signaling awaits additional analyses, including determining the molecular identity of the disrupted gene.

Emulsion regulates cytoplasmic segregation but not cortical granule exocytosis or other aspects of egg activation in zebrafish (Dosch et al., 2004). Compromised cytoplasmic segregation in *emulsion* mutants causes defective cellularization and developmental arrest of most progeny of mutant females (Dosch et al., 2004). Another maternal-effect gene, *Claustro*, is essential for chorion elevation, but is dispensable for cytoplasmic segregation during egg activation in the zebrafish

(Pelegri et al., 2004). Claustro may regulate chorion formation in the ovary (e.g., in mutants, the chorion may not be competent to respond to activation stimuli). Alternatively, claustro could be a target of, or one of the enzymes, released from the cortical granules upon egg activation to promote elevation of the chorion. Detailed analysis of the mutant ovaries and determining the molecular identity of the disrupted genes should provide insight into the mechanisms of claustro and emulsion action.

Together, the zebrafish maternal-effect mutants disrupting genes regulating multiple or unique components of egg activation demonstrate that the events of egg activation can be genetically uncoupled (Figure 11). Thus, egg activation is not an all-or-none phenomenon. The zebrafish mutants with phenotypes impacting all or several aspects of egg activation (e.g., cortical granule exocytosis, chorion elevation, cytoplasmic segregation) likely disrupt genes that act to initiate or coordinate egg activation events. Mutations disrupting only a subset of processes during egg activation may be hypomorphic alleles or represent genes with a dedicated function (for example regulating cortical granule exocytosis) for regulation of that component of egg activation. This collection of mutants will provide valuable insight into the molecular control of egg activation and the mechanisms that coordinate independent aspects of egg activation.

• • • •

Blocking Polyspermy

ROLES FOR POLARIZED SOMATIC FOLLICLE CELL TYPES

The sperm entry point is spatially restricted in amphibians, rodents, and zebrafish (Hart and Donovan, 1983; Johnson et al., 1975; Motosugi et al., 2006; Nicosia et al., 1977; Yanagimachi, 1994). In the frog sperm, entry is limited to the animal pole (Elinson, 1975). In mouse and hamster, the region overlying the meiotic spindle (the actin cap discussed earlier), comprising about 30% of the egg surface at the animal pole, is devoid of microvilli, and sperm-egg fusion is excluded in this region (Johnson et al., 1975; Nicosia et al., 1977; Yanagimachi, 1994) (Figure 3). In zebrafish, fertilization is restricted to a specialized channel, the micropyle, formed by a somatic follicle cell at the animal pole (Hart and Donovan, 1983) (Figure 3). In zebrafish, Bucky ball, discussed earlier for its role in regulating oocyte polarity, is also required to limit micropyle numbers during oogenesis to prevent polyspermy during fertilization (Marlow and Mullins, 2008). Bucky ball is not required for cortical granule exocytosis or chorion elevation, but these cortical events are not sufficient to block polyspermy due to the excess number of micropyles (Marlow and Mullins, 2008).

Polarized follicle cells are not unique to the zebrafish. In the mouse, the follicle cells also show polarized fates. In antral follicles (i.e., the small resting follicles with potential to be induced to mature), follicle-stimulating hormone (FSH) promotes antral cavity formation and contributes to the partitioning of the granulosa cells into mural or cumulus populations depending on their proximity to the oocyte and their distinct expression profiles (Figures 3 and 15). In the mouse, FSH and FSH receptor are required for development prior to the antral stage (Dierich et al., 1998; Kumar et al., 1997). FSH promotes development of the mural cells that line the follicle walls and contribute significantly to hormone production (Diaz et al., 2007). Initially, all granulosa cells express the luteinizing hormone receptor, *lhcgr.* Later a signal from the oocyte blocks *lhcgr* in the cells that are closest to the oocyte, the cells which develop as the cumulus cells, while *lhcgr* expression remains high in the cells further from the oocyte, the mural cells (Diaz et al., 2007). Evidence that Gdf9 and BMP15 are the signals from the oocyte that promote cumulus cell number, and fate by a SMAD2/3-dependent mechanism comes from the findings that the expression of *egfr* in cumulus cells is reduced when cultured without oocytes, in *gdf9* or *bmp15* individual and compound mutants, or in the presence of SMAD2/3 inhibitors (Diaz et al., 2007; Eppig, 2001; Hussein et al., 2006; Su,

Wu et al., 2004; Yan et al., 2001; Su et al., 2010). Each of these conditions compromises the ability of cumulus cells to respond to LH-induced EGF factors produced by the theca and mural cells. This cumulus–oocyte complex signaling network in turn generates positive feedback promoting maturation of the follicle, oocyte development, and ultimately, promotes the events that prepare the oocyte for ovulation and fertilization (Figure 15).

The asymmetric fluid filled cavity and signaling from the oocyte contribute to the formation of distinct follicle cell types relative to oocyte position and together the oocyte and follicle cells mediate formation of the zona pellucida. However, it is unclear whether these asymmetries in follicle cell type bare any relationship to the oocyte axis (e.g., do follicle cell asymmetries mark the future amicrovillar site of the oocyte or otherwise anticipate the region of the cortex where the nucleus will localize in late-stage oocytes?). The restricted fertilization site and apposition of pronuclei have been implicated in orienting the cleavage planes in early mouse embryos (Gardner, 2007; Motosugi

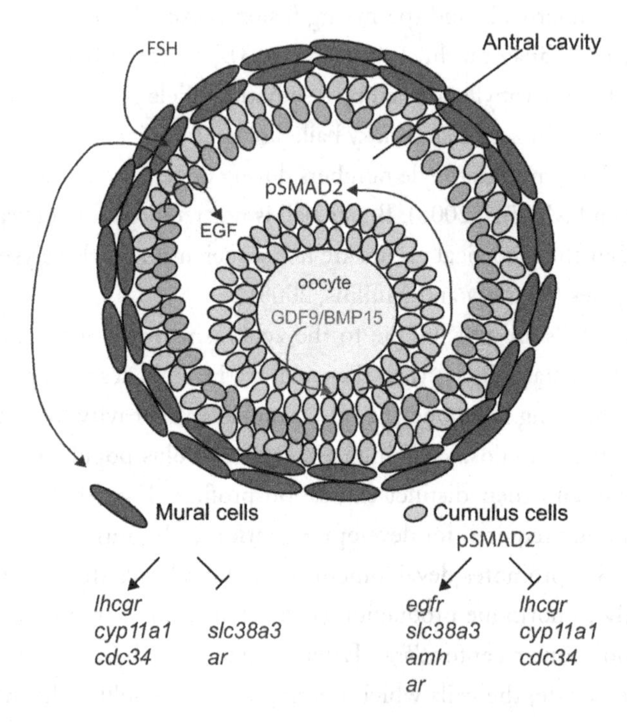

FIGURE 15: Model summarizing the signaling between the oocyte and the distinct somatic cell types that comprise the mammalian follicle. The cells closest to the oocyte show opposite expression profiles while the cells in between show intermediate versions of these extremes resulting in a well-patterned follicle.

et al., 2006; Piotrowska and Zernicka-Goetz, 2001; Plusa et al., 2005; Plusa et al., 2002; Zernicka-Goetz, 2002).

Together, the somatic cells, including the asymmetric granulosa cell populations and the oocyte, sustain development of the individual ovarian follicle. Maternal-effect mutants and conditional knockdown should facilitate further examination of the individual functions and contributions of these distinct asymmetrically positioned follicle cell types and how they influence the oocyte, fertility, and development of the prospective embryo.

UCHL1: A ROLE FOR DEUBIQUITINATION IN BLOCKING POLYSPERMY

The deubiquitinating enzyme, ubiquitin carboxy-terminal esterase L1 (UCHL1) protein, functions as both a hydrolase to remove and recycle monoubiquitin from degraded proteins and as a ligase to assemble ubiquitin tags to mark proteins for disposal. In mouse, UCHL1 protein localizes to the plasma membrane of oocytes and early embryos (Sekiguchi et al., 2006). The ovaries of *UCHL1* mutant mice are histologically normal and the mutant oocytes accomplish meiosis as judged by polar body extrusion, nonetheless, these mutant females produce fewer progeny (Sekiguchi et al., 2006). Multiple pronuclei in fertilized eggs lacking maternal *UCHL1* despite a normal cortical reaction based on analysis of zona pellucida proteins and cortical granule exocytosis provided evidence that *UCHL1* is required to block polyspermy (Sekiguchi et al., 2006).

Previous work by Sutovsky et al. (2004) demonstrated that zona pellucida proteins are ubiquitinated during oogenesis. Consistent with reports that ubiquitination occurs during oogenesis, Sekiguchi et al. (2006) observed an ubiquitination deficiency in UCHL1 mutant oocytes. Thus, UCHL1 mutants reveal a requirement for the ubiquitin–proteasome system (UPS) in preventing polyspermy. This was not the first time that UPS had been implicated in contributing to fertilization. Inhibition of the sperm acrosome associated proteosome function with inhibitors of UPS impeded sperm entry into zona intact mammalian oocytes and provided evidence that UPS degrades the sperm receptor from the zona surface (Sutovsky et al., 2004). At first glance, it seems that the results of these studies are contradictory (e.g., in one case, a decline of proteosome function via application of inhibitors blocks sperm entry while a decline in UPS function in mouse *UCHL1* mutants leads to an increase in sperm entry). However, when considering the data from the mammalian systems together, a potential model to reconcile this apparent discrepancy emerges, whereby UPS function delivered by the sperm upon acrosome binding leads to degradation of the ubiquitinated proteins on the surface. This would promote zona penetration and trigger UPS-mediated removal of the sperm receptor from the cell surface to prevent polyspermy. In *UCHL1* mutants the sperm receptor would be predicted to lack ubiquitin tags either due to the decreased hydrolase or ligase activity of

UCHL1. Thus, the receptors would remain on the surface and would be available to bind additional sperm despite an otherwise normal cortical reaction in the absence of UCHL1.

BASONUCLIN: A ZINC FINGER PROTEIN INVOLVED IN DEVELOPMENTAL COMPETENCE AND LIMITING SPERM ENTRY

Basonuclin is a zinc finger protein that associates with and promotes RNA polymerase II- and I-mediated transcription in the oocytes of mice (Tian et al., 2001; Tseng et al., 1999). Basonuclin depletion from mouse oocytes causes abnormal oocyte morphology and compromises female fertility (Ma et al., 2006b). Fertilized embryos show polyspermy and a range of defects likely attributable to abnormal cell cycle control including defective chromatin decondensation, defective DNA replication, and chromosome separation during the first cell division (Ma et al., 2006b). The morphological defects in oocytes and later failure to sustain embryonic development indicate that basonuclin may have essential functions at multiple developmental stages. However, additional studies are needed to demonstrate whether the oocyte defects cause the embryonic phenotypes or basonuclin has distinct functions in the oocyte and later in the embryo.

Intriguingly, some transcripts are more abundant in basonuclin-depleted oocytes, including Gli2, the transcription factor that promotes basonuclin expression (Ma et al., 2006b). This finding suggests that the oocytes can detect the deficiency in basonuclin levels and attempt to compensate. Alternatively, basonuclin could feedback to repress Gli2. The mechanism for detecting and responding to decreased Basonuclin is not known nor is the mechanism by which basonuclin functions to promote normal oocyte and early embryonic development understood. It is possible that aspects of the basonuclin depletion phenotype may be due to elevated expression of regulatory genes, such as Gli2, and their targets. Conditional approaches to perturb Hedgehog signaling in the gonad and uterus of the mouse underscore the need to precisely control Hedgehog activity for optimal fertility. Conditional knockdown of Indian Hedgehog in uterine endometrial cells blocks implantation and causes sterility (Franco et al., 2010). While a constitutively active Smoothened causes infertility due to impaired implantation, fertilization, and uterine development when expressed in the granulosa cells, or leads failure to release the oocyte and infertility when expressed under the control of anti-Müllerian hormone receptor 2 (Amhr2) (Ren et al., 2009; Franco et al., 2010). While these conditional approaches cannot reveal the full scope of Hedgehog contribution, these studies do indicate that excess Hedgehog signaling compromises fertility. Simultaneous knockdown of Gli2 and Basonuclin may help to distinguish between Gli2-dependent and any Gli2-independent aspects of the Basonuclin depletion phenotype.

• • • •

Cleavage/Mitosis: Going Multicellular

So far, we have discussed maternal-effect genes that are necessary to produce a fertilization competent egg. Once fertilized, the maternal and paternal genetic material (the pronuclei) will fuse, and the now single-celled embryo must divide to produce the cells required to form the multicellular animal. While the precise number varies depending on the animal, the first several mitotic cell cycles occur before the zygotic or embryonic genome is expressed; thus, these cell cleavages depend on maternal products as well as contribution from the fertilizing sperm (Kim and Roy, 2006; Yabe et al., 2007; Zhong et al., 2005). The maternal factors involved are largely unknown but are expected to include proteins that mediate the destruction of meiotic factors and post-translational regulation of mitotic cell cycle regulators. In this chapter, we will discuss the maternal genes required for the early cell divisions of the embryo.

In some vertebrates, such as humans and mouse, the mitotic cycle begins prior to or concomitant with maternal and paternal pronuclei fusion, while in other animals (e.g., zebrafish, *Xenopus*), the pronuclei fuse before mitosis. In zebrafish, mitotic cycle initiation does not absolutely require pronuclear fusion as is evident from *futile cycle* maternal-effect mutants in which haploid nuclei fusion is blocked; nonetheless, cytokinesis persists (Dekens et al., 2003). The cleavages of early vertebrate embryos are distinct from the cell divisions that occur later in development. The early embryonic cleavages are more rapid, and the cycles consist of synthesis and division phases without gap or intervening phases (Figure 16). Consequently, each cell division amplifies the total number of cells and the DNA content, but the resulting cells become increasingly smaller with each division during the cleavage phase. In animals that develop rapidly, more cleavage divisions are completed before the zygotic genome is activated (e.g., 12 synchronous cleavage cycles in *Xenopus* and zebrafish; Newport and Kirschner, 1982; Kane and Kimmel, 1993) and 4 in bovine embryos]. In these animals, there is a prolonged temporal window separating regulation of cellular cleavages and zygotic genome activation. Thus, the genes that are essential to regulate early cleavages can be distinguished from those genes that are not required for cleavage per se, but instead are necessary for zygotic or embryonic genome activation (ZGA, EGA). At or after zygotic genome activation, the cell cycle lengthens as gap, and intervening phases are introduced between synthesis and division (Figure 16).

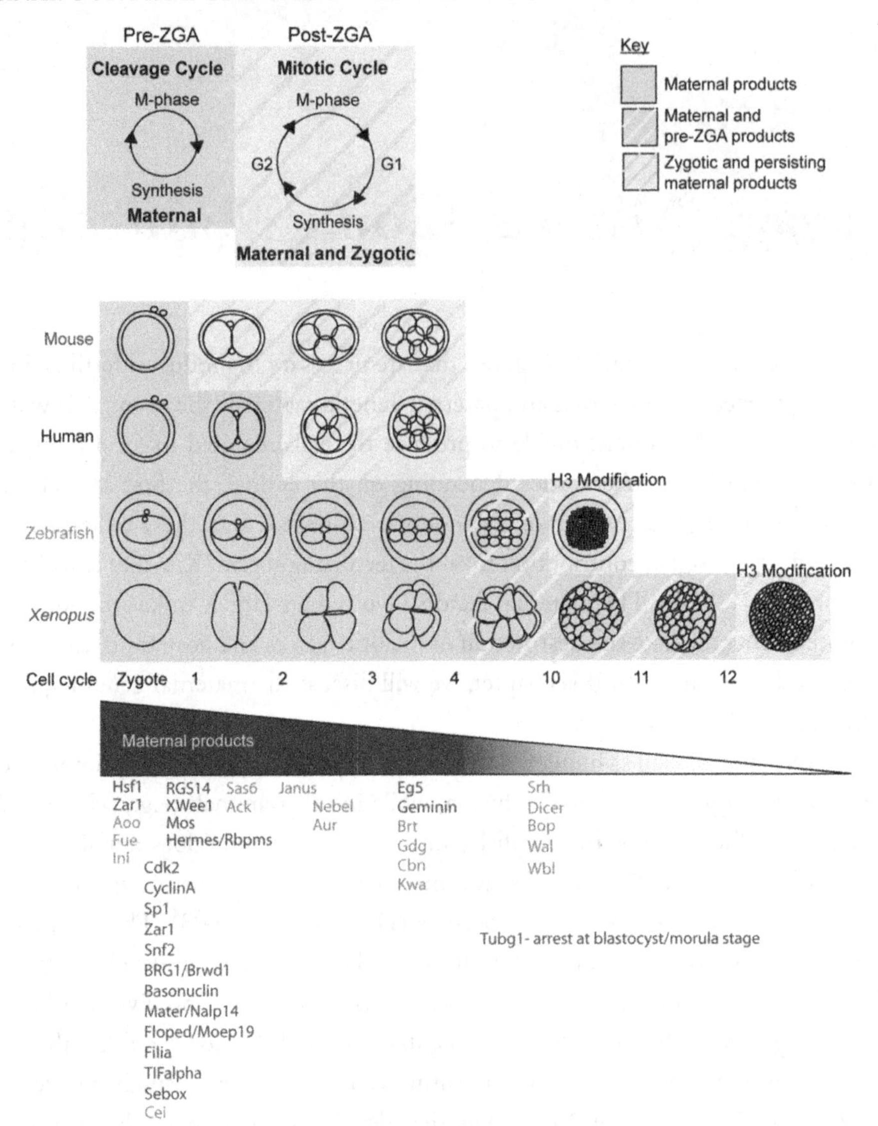

FIGURE 16: Schematic depicting the maternal to zygotic transition. The cleavages of early embryos lack gap phases, are rapid, generate increasingly smaller cells, and are under maternal control. Upon zygotic genome activation, the cell cycle lengthens as gap phases and asynchronous divisions are introduced. These mitotic divisions are under maternal and zygotic control in some animals. In the mouse and humans, the cell cycles are longer and zygotic contribution begins earlier compared with zebrafish and *Xenopus*. The genes known to be required for early cleavage or zygotic genome activation are listed according to the stage when their maternal function is required for each model organism.

GENETIC UNCOUPLING OF KARYOKINESIS AND CYTOKINESIS IN ZEBRAFISH

Cell division or cleavage requires duplication of the chromosomes and centrioles followed by equal partitioning of the genetic material (karyokinesis) and division of the cytoplasm (cytokinesis). Distinct classes of maternal-effect mutants in which karyokinesis and cytokinesis are genetically uncoupled have been identified through genetic screens (Figure 16) (Dosch et al., 2004; Kishimoto et al., 2004; Pelegri et al., 2004; Pelegri et al., 1999). One class of zebrafish maternal-effect mutants disrupt genes required to initiate cleavage; *indivisible* and *atomos* mutants are fertilized, but do not divide (Dosch et al., 2004). In these mutants, both karyokinesis and cytokinesis are blocked (Dosch et al., 2004). The spindle and the nuclear envelope have not been examined nor has the possibility that defective pronuclear fusion contributes to these mutant phenotypes been ruled out. However, the zebrafish *futile cycle* mutants discussed above are defective in pronuclear fusion yet undergo cytokinesis indicating that fusion of the maternal and paternal genetic material is not prerequisite for cytokinesis, but is essential for karyokinesis (Dekens et al., 2003).

In the phenotypic class comprised of *golden gate* (*gdg*), *kwai*, *bo peep*, and *waldo* mutants, cytokinesis proceeds despite impaired karyokinesis (Pelegri et al., 2004). In contrast, mitosis occurs without cytokinesis in *ack* mutants (Kishimoto et al., 2004). Finally, *weeble* and *barrette* are required for both cytokinesis and mitosis, as is *cobblestone*, but only in a subset of cells (Pelegri et al., 2004). This collection of mutants disrupts genes that initiate or coordinate both karyokinesis and cytokinesis and provide genetic evidence for distinct regulation of the nuclear and cellular division aspects of early cleavages. Although the molecular identities of the disrupted genes required for early cleavage have yet to be determined, it is possible that some of the genes encode regulators of meiotic factor clearance or mitosis activators.

CONTRIBUTIONS OF THE CYTOSKELETON TO CYTOKINESIS AND KARYOKINESIS

Evidence that the cytoskeleton sustains cleavages comes from interference studies using antibodies and cytoskeletal inhibitors of actin, tubulin, and cytokeratin during embryonic cleavage stages in model systems, including zebrafish, frog, and mouse. These studies uncovered maternal contributions of the cytoskeleton to regulation of the spindle, of furrow formation, of endocytosis, of membrane remodeling, and of cellular cohesiveness via delivery of adhesion molecules to the cell surface in cleavage stage embryos (Danilchik et al., 1998; Feng et al., 2002; Jesuthasan, 1998; Zhong et al., 2005). More recently, zebrafish maternal-effect screens have contributed genetic support for essential and distinct contributions of cytoskeletal components to controlling the early cleavages.

Depending on the cell, cleavage furrows form at the cortex in the vicinity of the spindle apparatus in response to signals emanating from the spindle itself or from the associated astral microtubules (Bringmann, 2008; Bringmann et al., 2007; Bringmann and Hyman, 2005; Bringmann, Skiniotis et al., 2004). The zebrafish maternal-effect *cellular island* and *futile cycle* mutants support a role for signals from both (Dekens et al., 2003; Yabe et al., 2009). *Cellular Island* encodes AuroraB Kinase, previously implicated in regulating furrow formation (reviewed by Ruchaud et al., 2007). In zebrafish, *cellular island* is required to form the astral microtubule-associated furrows, but not the spindle-induced furrows (Yabe et al., 2009). Microtubule nucleation and spindle assembly require *futile cycle* (*fue*) function while furrow formation and cell partitioning are independently regulated in the zebrafish (Dekens et al., 2003). Although the spindles are not properly formed in *fue* mutants, those spindles that are intact are sufficient to initiate furrow formation (Dekens et al., 2003). Thus, both mechanisms seem to contribute nonredundantly in the zebrafish to regulation of distinct subsets of the cleavage furrows present in the early embryo.

After furrow initiation, the product of the zebrafish *nebel* gene is required for cleavage furrow microtubule array formation (Pelegri et al., 1999). Mutants disrupting the zebrafish *aura* gene indicate that aura promotes membrane recruitment to the furrow in order to accomplish cleavage by a mechanism that is not understood (Pelegri et al., 2004). Regulation of actin dynamics during cleavage furrow formation is mediated by *acytokinesis* function, which is required to accomplish successful cytokinesis and karyokinesis during early cleavage stages in zebrafish embryos (Kishimoto et al., 2004). Determining the molecular identity of these disrupted genes promises to provide insight into the molecular mechanisms regulating microtubule and actin dynamics during separation of the chromosomes and cellular partitioning.

CELLULAR ATOLL/SAS6: CENTROSOME DUPLICATION AND KARYOKINESIS IN ZEBRAFISH

Centrosomes contribute to preservation of ploidy through their function in organizing the spindle poles during cell division. The eggs of many animals are devoid of centrioles until fertilization when the sperm provides the centrioles necessary for development of the embryo. The centrioles are duplicated prior to the first cell division to generate the centrosomes that will mediate chromosome separation during cell division. The zebrafish *cellular atoll* mutant disrupts a conserved residue within the centriolar protein Spindle assembly 6, Sas6. Evidence for an essential maternal role for Sas6 in formation and duplication of centrosomes during embryonic cleavages comes from a hypomorphic allele disrupting *Sas6* (Yabe et al., 2007). The first cell division is normal in the progeny of mutant mothers; however, during the second division cycle, a subset of *cellular atoll* progeny develops with a monopolar spindle and a single centriolar pair (Yabe et al., 2007). The consequence of failed centrosome duplication during the second cell division cycle is defective karyokinesis and furrow formation (Yabe et al., 2007). Not all animals rely on centrosome-dependent spindle assembly during early cleavages, notably, in the mouse, centriolar assembly is regulated independent of early cleavage (Wadsworth and Khodjakov, 2004).

Pms2: A ROLE FOR MISMATCH REPAIR IN PREVENTING ANEUPLOIDY

The DNA mismatch repair pathway corrects base pair errors and small deletions or insertions. During the initial rapid cleavages of mouse embryos maternal post-meiotic segregation increased 2 (Pms2), a functional homolog of the mismatch repair protein MutL, functions to limit the accumulation of mutations during replication as is evidenced by the replication errors that accumulate in embryos lacking maternal Pms2 at the one-cell stage (Gurtu et al., 2002). The *pms2* maternal-effect mutant phenotype indicates that, even after the zygotic genome has been activated, the embryo seems to have no mechanism to remedy the aneuploidy caused in the absence of maternal PMS2 function, while surviving *Fmn2* mutants discussed earlier indicate that a mechanism may exist to correct aneuploidy (Dumont et al., 2007; Gurtu et al., 2002; Leader et al., 2002). The reasons for the differences in ability of zygotic gene function to compensate or correct the error may lie in the developmental timing, meiosis versus first cleavage. Alternatively, it may be that an excess of chromosomes, as occurs in *Fmn2* maternal mutants can be repaired, while the widespread deletions, insertions, and rearrangements causing inappropriate copy numbers observed in MutL mutants are beyond the capacity of the repair machinery.

GEMININ: LIMITING REPLICATION WITH LICENSING COMPONENTS

Geminin is a component of the licensing machinery that acts to limit DNA replication to a single round during each cell division cycle. Geminin deficiency in the mouse results in developmental arrest at the four- to eight-cell stage when maternally supplied Geminin protein is thought to be exhausted or degraded (Hara et al., 2006; Gonzalez et al., 2006). Based on the correlation between degradation of the maternal protein, the onset of the SI-phase block, and ensuing excessive endoreplication, which causes premature differentiation of trophoblast cells in zygotic *Geminin* mutants, maternal Geminin is predicted to prevent the cells of the early embryo from adopting trophoblast fate at the expense of the pluripotent ICM cells until zygotic Geminin protein is produced (Hara et al., 2006; Gonzalez et al., 2006). A maternal mutant disrupting Geminin is needed to definitively demonstrate an essential role for maternal Geminin in preventing precocious trophoblast differentiation.

MICROTUBULES AND MOTORS: REGULATING THE SPINDLE AND POLAR BODY

A plus-ended kinesin-related motor protein, Eg5, regulates proliferation during early cleavage stages in the mouse (Castillo and Justice, 2007). Although not directly examined in the Eg5 knockout, the proliferation defects have been ascribed to Eg5 function in spindle assembly. Maternal *eg5* mRNA and protein peaks in the oocytes of mouse and *Xenopus* are consistent with a maternal role for Eg5; this maternal contribution has been proposed to sustain normal development of zygotic mutants until the maternal Eg5 protein stores become limiting at zygotic genome activation (Houliston et al., 1994; Sawin and Mitchison, 1995; Winston et al., 2000; Zeng and Schultz, 2003).

Mouse γ-tubulin is also essential for spindle formation and proper nuclear division. In γ-tubulin mutants, defects in nuclear division correlate with, and thus have been attributed to the temporal exhaustion of maternal γ-tubulin protein stores (Yuba-Kubo et al., 2005). Depletion and gain of function studies in oocytes support a maternal contribution of γ-tubulin to spindle regulation. Depleting γ-tubulin diminishes spindle and polar body size, while supplying an excess of maternal γ-tubulin during meiotic divisions increases spindle and polar body size (Barrett and Albertini, 2007). Interestingly, polar bodies that receive additional γ-tubulin are not only larger, but also undergo cytokinesis rather than degradation after they are extruded (Barrett and Albertini, 2007). How the total pool of γ-tubulin available to the oocyte is partitioned to ensure that sufficient γ-tubulin is allocated to complete meiosis, yet enough is reserved to permit normal progression through the early cleavage cycles of the embryo is not understood.

Tcl1: CONTRIBUTION OF A CELL SURVIVAL FACTOR TO EMBRYONIC CLEAVAGES

T cell leukemia/lymphoma 1, *tcl1*, encodes a protein with globular domains that functions as a cofactor of Akt1 in cell culture. *Tcl1* knockout mice develop as fertile males or as females with compromised fertility, but maternal Tcl1 is not essential for meiosis or fertilization (Narducci et al., 2002). When cultured, the progeny of mutant mothers fail to develop to blastocyst stages; arrest of the majority of embryos at the four- to eight-cell stage indicates a necessary role for maternal Tcl1 during preimplantation stages (Narducci et al., 2002). Narducci and colleagues showed that Tcl1 protein shuttles between the cell cortex and the nucleus through the eight-cell stage of development in a cell cycle-dependent manner. Consistent with the cell cycle-regulated shuttling of the protein, maternal Tcl1 is required for cleavage (Narducci et al., 2002). Normal expression of zygotic genome activation reporters and initiation of compaction in the progeny of Tcl1 maternal mutants provided evidence that zygotic genome activation and differentiation are Tcl1-independent processes. Though dispensable for compaction, reduced numbers of cells were observed in the progeny of maternal Tcl1 mutants (Narducci et al., 2002). Therefore, it is possible that the embryos arrest because they lack sufficient cells to form a proper inner cell mass. However, the mechanism by which maternal Tcl1 regulates early embryonic cleavages and the transition to the blastocyst stage remains to be elucidated.

Most candidate genes postulated to function maternally during early cleavage stages were chosen based on their expression profiles. For example, transcripts with robust expression at, or prior to, the first cleavages and subsequent rapid elimination thereafter are candidate regulators of early cleavages or of embryonic genome activation (EGA). In mouse, the maternal to zygotic transition occurs after only one cell cycle (Figure 16) (Mager et al., 2006; Mathavan et al., 2005; Melton, 1991; Memili and First, 2000; Misirlioglu et al., 2006; Zimmermann and Schultz, 1994). In general, embryos that activate the zygotic genome during the initial cleavage stages would be expected to arrest at a similarly early developmental stage when maternal function is depleted. Accordingly, in the mouse, many maternal-effect mutants arrest at around the two-cell stage. The switch from cleavage to mitosis cycles and from maternal to zygotic control of development both occur around the time of the first cell cycle. Therefore, in the mouse, mutations that disrupt genes required for mitosis or genes required for zygotic genome activation will cause developmental arrest at the two-cell stage and fail to implant (Figure 16). On the other hand, mutations that disrupt mitosis regulators may not show a phenotype due to compensation by the zygotic counterpart. If an embryo does arrest before EGA, it is not likely to express zygotic genes. Conversely, if the zygotic genome is not activated, the ability of the embryo to continue to cleave will be compromised. The temporal proximity of these developmental events makes it extremely challenging to distinguish whether a gene has a specific and essential role in regulating the cell cycle, other aspects of the first embryonic cleavages, or is required to activate the embryonic genome.

• • • •

Maternal–Zygotic Transition

In the previous chapter, we discussed the known maternal proteins required for early cleavages and initial cell divisions of the fertilized vertebrate embryo. The switch from the cleavage type (i.e., no gap phases) to mitosis with gap phases involves activation of zygotic gene expression to support the subsequent mitotic cell divisions. In this chapter, we will review the essential maternal regulators mediating elimination of maternal products and activation of zygotic gene expression. The maternal to zygotic transition is not a single event, but rather, a series of events. This transition begins with activation of the egg and extends through activation of zygotic gene expression and culminates with the complete elimination of maternal products. The extent of the turnover of maternal products and the developmental stage when maternal clearance initiates and concludes varies. In some animals, maternal products are not completely eliminated until stages well after zygotic gene expression begins. Despite this variation, the switch from full maternal control to zygotic gene activation is initially directed by maternal products, which act to promote more robust transcription of zygotic genes or to degrade maternal products during zygotic genome activation.

PROVIDING AMINO ACIDS VIA AUTOPHAGY

In the mouse, fertilization activates the egg and promotes the transition from oocyte to embryo. At this time, maternal mRNAs and proteins that were important for oogenesis, but are not essential for embryogenesis, are degraded. Two key cellular systems regulating protein degradation are the ubiquitin proteasome and macroautophagy. In mammalian eggs, fertilization activates autophagy as is evident from association of Atg-8 (autophagy related 8)/LC3-GFP, a reporter for autophagy, with the lysosomes of fertilized eggs but not unfertilized eggs (Tsukamoto et al., 2008a; Tsukamoto et al., 2008b). Decreased phosphorylation of S6 kinase (an indication of reduced activity of the negative regulator of autophagy mTOR) in fertilized eggs provided additional evidence that autophagy increases after fertilization (Tsukamoto et al., 2008a; Tsukamoto et al., 2008b). Oocyte-specific knockout of Autophagy related 5 (Atg-5) resulted in embryonic arrest prior to blastocyst stage, a defect that can be partially compensated by zygotic Atg-5, and the resulting decline in ^{35}S incorporation without diminished protein synthesis rates provide functional evidence that macroautophagy-mediated protein degradation provides amino acids necessary for protein synthesis during preimplantation development of the mouse embryo (Tsukamoto et al., 2008a; Tsukamoto et al., 2008b). Whether autophagy similarly contributes to post-fertilization development in other vertebrates is not known.

A small number of zygotic transcripts have also been detected before Zygotic Genome Activation (ZGA) including β-catenin and XTcf targets (e.g., *bozozok* in zebrafish, *nodal-related 5* and *nodal-related 6* in *Xenopus*) (Yang et al., 2002; Leung et al., 2003). Apart from such exceptions, fertilized eggs and early embryos are largely transcriptionally silent until the ZGA. Consequently the maternal products present in fertilized eggs are regulated predominantly by post-transcriptional means. Elongation of poly (A) tails is crucial for mRNA stability and translational control in early embryos. In mature oocytes and following fertilization in frogs and mouse polyadenylated maternal mRNAs are selectively recruited for translation (Barkoff et al., 2000; Fuchimoto et al., 2001; Paynton and Bachvarova, 1994). In the mouse, maternal cyclinA is essential for zygotic genome activation and is activated by Cdk2-mediated phosphorylation of Retinoblastoma, Rb (Hara et al., 2005), and possibly also by Sp1 transcription factor (Fojas de Borja, Collins et al. 2001) mediated transcription.

Considerable effort toward understanding the switch from maternal to zygotic control thus far has focused on transcript analysis or changes in gene expression at stages prior to and after ZGA (Clegg and Piko, 1982; Ferg et al., 2007; Mathavan et al., 2005; Piko and Clegg, 1982; Alizadeh et al., 2005). The overall absence of detectable *de novo* transcription and the abundant maternal stores present during early maternally controlled development suggest that much of the regulation is likely translational and post-translational. Analysis of complimentary data sets from multiple approaches including proteomics, epigenome analyses, and systematic assessment of post-translational modifications (e.g., phosphorylation state) occurring at stages prior to and after ZGA may provide insight into how activation of the zygotic genome is regulated. However, a full understanding of the maternal to zygotic transition will require identification of a mutation or a collection of mutants disrupting the ZGA. So far, a "master regulator" of ZGA has not been identified in vertebrates.

miRNAs: SILENCING MATERNAL MESSAGES

MicroRNAs (miRNAs) and short interfering RNAs (siRNAs) have recently become appreciated as key regulators of post-transcriptional regulation and mRNA turnover. miRNAs are produced from double-stranded RNAs. The primary transcript or pri-miRNA is processed into pre-mRNA hairpins in the nucleus by the microprocessor complex composed of the double stranded RNA binding protein DGCR8, also known as Pasha, and the RNAse III family member, Drosha. The pre-mRNA stem loop structure is then exported to the cytoplasm where it is recognized and cleaved by another RNAse III, *Dicer*, to generate the mature 22 nucleotides in length miRNA (Knight and Bass, 2001; Bernstein et al., 2001; Grishok et al., 2001; Hutvagner et al. 2001; Ketting et al., 2001; Hammond et al., 2000; Hutvagner and Zamore 2002). The mature miRNA promotes RNA-induced silencing complex (RISC)-mediated translational repression or nuclease-mediated cleavage and degradation of the endogenous complimentary messenger RNA (recently reviewed by Hutvagner and Simard, 2008). Studies in vertebrate model systems provide evidence that small RNA biogenesis contributes to mRNA turnover during developmental transitions and differentiation.

In zebrafish, selective degradation of maternal mRNAs, rather than wholesale turnover of maternal mRNAs, occurs at the midblastula transition (MBT) or upon ZGA. Compromised deadenylation and decay of maternal transcripts in zebrafish maternal-zygotic *dicer* mutants support involvement of *dicer* in clearing maternal messages (Figure 17) (Giraldez et al., 2005). The expression of zygotic genes in maternal-zygotic *dicer* mutants demonstrated that the ZGA is initiated independent of Dicer-mediated clearance of maternal mRNAs, and provided evidence that eliminating maternal messages is not prerequisite or essential for wholesale zygotic genome activation (Giraldez et al., 2005; Giraldez et al., 2006).

Maternal-zygotic *dicer* mutants display profound gastrulation and embryologic dysmorphology suggesting that clearance of maternal transcripts is essential for normal development, but not for specification or patterning of embryonic tissues (Giraldez et al., 2005; Giraldez et al., 2006). Suppression of the early phenotypes, but not the later lethality of *MZdicer* mutants when miR-430 is supplied to the embryo indicates that the miR-430 class is largely responsible for mediating clearance of maternal messages soon after activation of the zygotic genome (Figure 17) (Giraldez et al., 2006). The relatively mild zygotic and maternal-zygotic *dicer* loss of function phenotypes led Giraldez and colleagues to propose that miRNAs sharpen the boundary between maternal and zygotic control of development (Giraldez et al., 2005; Giraldez et al., 2006) and reviewed by Schier and Giraldez (2006). In zebrafish maternal miRNAs have not been detected (Schier and Giraldez, 2006). Upon ZGA, miR-430 is processed by Dicer; thus, miR-430 activation couples the clearance of maternal mRNAs to activation of the zygotic genome (Figure 17). This appears to be a conserved mechanism to eliminate maternal messages as the *Xenopus* ortholog, miR-427, is expressed several hours before wholesale zygotic genome activation and regulates deadenylation and decay of maternal messages, including *cyclins* (Lund et al., 2009).

In the mouse, *dicer* also regulates mRNA clearance during development, including at maternally regulated transitions as is evident from the phenotypes of tissue-specific knockouts. In oocytes, dicer regulates mRNA turnover associated with maturation; *zona pellucida Cre* (*ZPCre*)-mediated knockdown disrupts spindle organization, chromosomal attachment and segregation, and extrusion of the polar body during meiosis I (Murchison et al., 2007). Consistent with the spindle defects, mRNAs regulating microtubule-related processes and mRNAs that decline in abundance upon maturation were among those more abundant in *dicer* mutant oocytes (Figure 17) (Murchison et al., 2007). Due to the early role of dicer in clearing mRNAs accompanying oocyte maturation, it

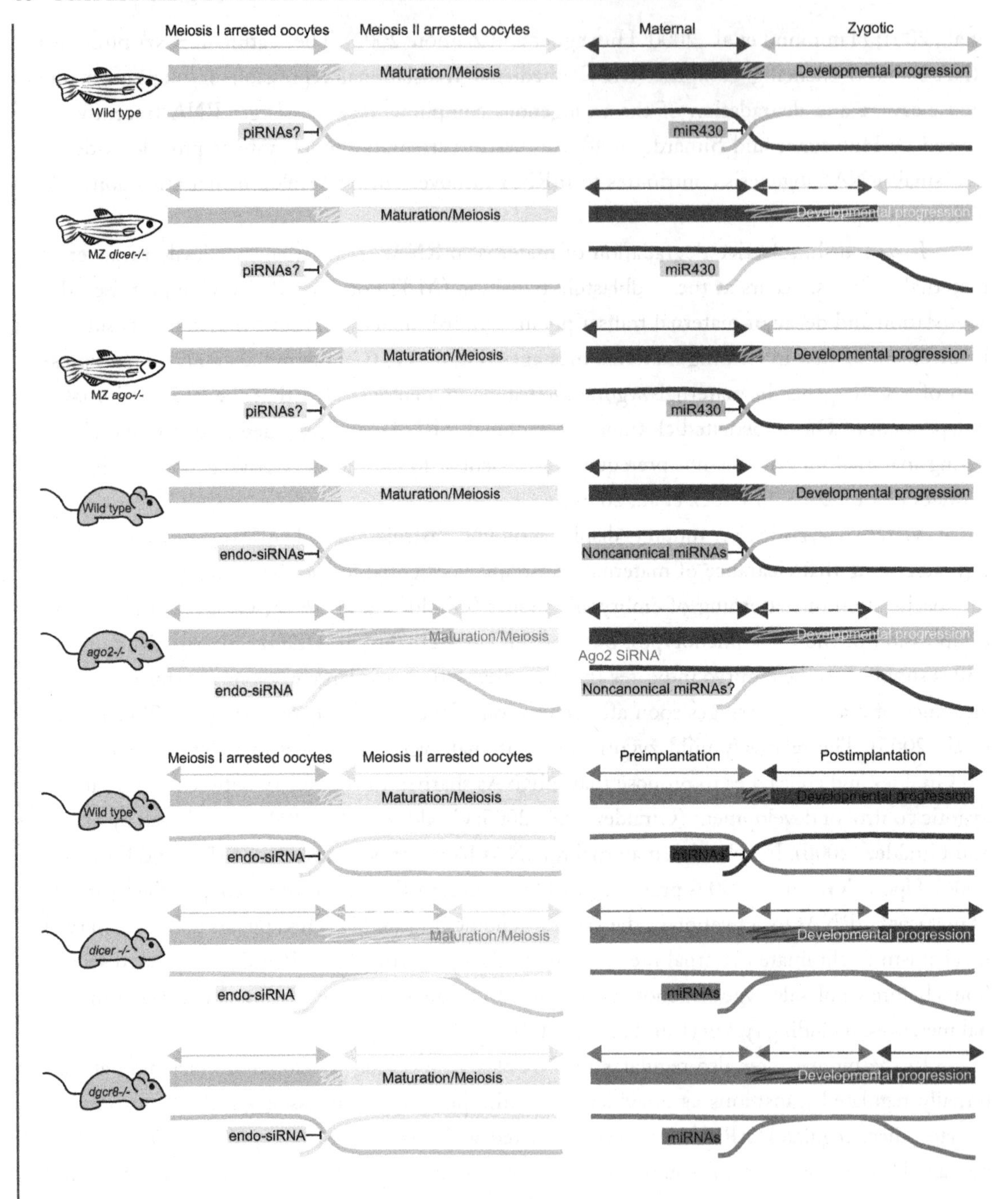

Meiosis I arrested oocytes
Meiosis II arrested oocytes
Maternal
Zygotic
Maturation/Meiosis
Developmental progression
Wild type
piRNAs?
miR430
MZ dicer-/-
MZ ago-/-
endo-siRNAs
Noncanonical miRNAs
ago2-/-
endo-siRNA
Ago2 SiRNA
Noncanonical miRNAs?
Preimplantation
Postimplantation
miRNAs
dicer -/-
dgcr8-/-

was not possible to examine its role at EGA. The oocyte maturation phenotype of the *ZPCre-dicer* knockout is more severe compared with the zebrafish *MZdicer* mutants, which produce normal oocytes (Giraldez et al., 2006). Similar to zebrafish *MZdicer* mutants, the maternal contribution of the double-stranded RNA binding protein *dgcr8* is not required for normal oogenesis, but is essential for embryonic development (Suh et al., 2010; Wang et al., 2007). Unlike Dicer, which processes both miRNAs and endo-siRNAs, Dgcr8 processes miRNAs but not endo-siRNAs. Thus, *dgcr8* mutants provide an opportunity to examine the consequences of loss of miRNAs independent from loss of endo-siRNAs. The requirement for endosiRNAs, but not miRNAs to regulating RNA decay during oocyte maturation is evident from the several hundred misregulated mRNAs in *dicer* mutants and normal mRNA profiles of *dgcr8* mutant oocytes (Figure 17) (Murchison et al., 2007; Suh et al., 2010; Wang et al., 2007). These findings demonstrated that *dicer* processed small RNAs uniquely contribute to mRNA clearance at distinct developmental transitions; endo-siRNAs are important for mRNA clearance during oocyte maturation, while miRNAs are crucial for clearance of mRNAs in the embryo.

Although Dicer is the key ribonuclease III enzyme responsible for processing miRNAs from primary transcripts, mature miRNAs are also produced by Dicer-independent mechanisms as indicated by the presence of mature miRNAs even in mouse ES cells in which *dicer* was conditionally ablated and in zebrafish MZ*dicer* mutants (Cheloufi et al., 2010; Cifuentes et al., 2010). Among the mature miRNAs produced in zebrafish MZ*dicer* mutants, miR-451 is processed by MZ*argonaut2* (*ago2*) slicer activity as evidenced by its presence in MZ*dicer* mutants and absence in zebrafish MZ*ago2* and mouse *ago2*-null mutants (Cheloufi et al., 2010; Cifuentes et al., 2010). Support for Ago2 function independent of Dicer comes from processing of miR-451 in *dicer* ablated mouse ES cells and rescue of the brain morphogenesis defects of zebrafish *MZdicer* mutants by a miR-430-*ago2*-hairpin, but not a dicer-dependent hairpin (Cheloufi et al., 2010; Cifuentes et al., 2010). Thus, maternal and zygotic *ago2* likely contribute to processing or stability of miR-451, and other similar hairpin-structured miRNAs by a dicer-independent mechanism in zebrafish. Maternal Ago2 is not

---

FIGURE 17: In the mouse, Dicer and Ago2 are required during oogenesis. Dicer processes endo-siRNAs (yellow) to promote mRNA (light blue) decay during maturation. When these mRNAs are not cleared, the oocytes arrest with defective meiosis. In zebrafish, the mechanisms acting to eliminate mRNAs after maturation do not require Dicer or Ago2, but may involve piRNAs. In zebrafish, Dicer processes miR430 at the time of zygotic genome activation, thus coupling maternal mRNA decay to the onset of zygotic gene expression. Ago2 depletion implicates noncanonical miRNAs (green) in degradation of maternal mRNAs (purple) to sustain development after zygotic genome activation. In the mouse, Dgcr8 processes miRNAs (red) that are required to degrade mRNAs (dark blue) and sustain development after implantation.

strictly required for development of zebrafish embryos as MZ*ago2* mutants are morphologically normal during gastrulation and early organogenesis (Cheloufi et al., 2010; Cifuentes et al., 2010). Compromised erythrocyte maturation due to failure to produce the erythroid-specific miR-451 miRNA in zebrafish MZ*ago2* and mouse *ago2*-null embryos indicates that dicer-independent Ago2 function in hematopoietic maturation is conserved.

In the mouse, *argonaute2* genes are expressed in oocytes. ZP3Cre-mediated knockdown of Ago2 revealed its contribution to mRNA turnover associated with maturation, as evident from the similarity between *ago2* oocyte knockdown and *dicer* mutant phenotypes (hundreds of misregulated mRNAs and meiotic spindle defects) (Kaneda et al., 2009). RNAi knockdown supports a later role for *ago2* as an essential mediator for development beyond the two-cell stage (Lykke-Andersen et al. 2008). As discussed earlier, elimination of maternal mRNAs and expression of zygotic transcripts are robust at the two-cell stage during the transition from maternal to zygotic control of development in the mouse. As anticipated based on its role in RISC-mediated silencing, a subset of maternal RNAs that should be degraded in two-cell stage mouse embryos were more abundant in *ago2*-depleted embryos (Lykke-Andersen et al., 2008). In contrast, some zygotically expressed transcripts were less abundant, indicating that failed decay of some maternal messages may adversely impact the expression of zygotic genes essential for developmental progression in the mouse. However, which Ago2-dependent miRNAs are required for developmental progression remains to be determined. These studies illustrate how post-transcriptional regulation mediated by distinct classes of small RNAs contributes to robustness both during global developmental transitions and in a cell- or tissue-specific manner during differentiation.

Zfp36l2: ZINC FINGER-MEDIATED CLEARANCE OF MATERNAL MESSAGES

Zinc finger protein 36 like 2, also known as TIS11D, ERF2, and BRF2, a CCCH tandem zinc finger protein, belongs to a family of zinc finger proteins that bind to AU rich elements within the 3′UTR of mRNAs and promote their degradation (Lai et al., 1999; De et al., 1999; Carballo et al., 1998; Blackshear, 2002; Lai et al., 2000). In the mouse, targeted disruption of *zfp36l2* produces a truncated Zfp36l2 protein; however, the resulting mutants are viable, and the mutant females have normal ovarian histology and meiosis as indicated by normal polar body extrusion (Ramos et al., 2004). The maternal-effect phenotypes of Zfp36l2 females revealed its maternal requirement for development beyond the two-cell stage, but how Zfp36l2 contributes to developmental progression is not known (Ramos et al., 2004). The function of related zinc finger family members in promoting transcript degradation point toward a potential role for Zfp36l2 in eliminating maternal mRNAs at the one–two cell transition when roughly 40% of the maternal mRNAs are degraded (Bachvarova and De Leon, 1980). However, RNA stability has not been examined in Zfp36l2 mutants. It will be particularly interesting to determine whether Zfp36l2 regulates turnover of a particular class of mRNAs, and if so whether persistence of those RNAs causes the maternal-effect arrest phenotypes.

A CLOCK MECHANISM REGULATES EGA IN THE MOUSE

The transcript profile of the mouse embryo undergoes three distinct transformations during early development; these occur at the zero- to two-cell stage, the four-cell stage, and during blastocyst differentiation (Hamatani 2004, Wang 2004). Both transcription and translation are repressed in the mouse embryo until early zygotic genome activation. In mouse, embryonic genome activation (EGA) is regulated by a timing mechanism, or a clock, that is triggered by fertilization rather than a mechanism coupled to the cell cycle. The basis for the clock or timer mechanism comes from cell cycle inhibitor studies where embryos are blocked prior to S-phase, yet still initiate the early zygotic genome activation program (Bolton et al., 1984; Schultz, 1993; Wiekowski et al., 1997; Aoki et al., 1997). The zygotic clock of mouse embryos is likely regulated by activation of a general transcription factor or alleviation of its repression based on observations that plasmids bearing RNA polymerases I, polII, and polIII promoter show comparable temporal patterns of activity when exogenously supplied during early zygotic genome activation (Majumder and DePamphilis, 1994). Several observations indicate that repression of the zygotic genome in early mouse embryos until the early EGA is mediated by factors derived from the oocyte. For example, DNA injected into the maternal pronucleus, but not the paternal pronucleus is repressed, and transcriptional activity of the paternal pronucleus is more robust compared to the maternal pronucleus (Ram and Schultz, 1993; Wiekowski et al., 1993; Aoki et al., 1997; Bouniol et al., 1995). Moreover, dispermic or trispermic mouse eggs are not more transcriptionally active; thus, the transcriptional capacity of the embryo is thought to be limiting (Worrad et al., 1994; Aoki et al., 1997). The female and male pronuclei have also been proposed to compete for the available pool of maternally supplied transcription factors, with the male pronucleus being more effective or accessible based on observations that the male pronucleus contains a higher concentration of transcription factors, and is more transcriptionally active prior to early EGA (Worrad et al., 1994). In support of this notion, when mouse eggs are parthenogenetically activated, the female pronucleus is transcriptionally more similar to the male pronucleus of monospermic eggs (Worrad et al., 1994; Aoki et al., 1997).

As alluded to in the previous chapter, the early cleavages of mouse embryos are relatively slow and take 12–24 hours. In the mouse activation of the zygotic genome occurs in two waves; the first wave occurs in the pronuclei at the one cell stage and the second takes place at the two-cell stage (Schultz, 2002; Thompson et al., 1998). Despite this rapid transition from maternal to zygotic control, the cell cycles occurring after the zygotic genome activation continue to resemble early cleavages in that they produce smaller cells with an increased nucleo-cytoplasmic ratio until implantation (Aiken et al., 2004). It is possible that the relatively long cell division cycle of the early mouse embryo necessitates early activation of zygotic gene expression. For example the maternal RNAs and proteins required for cell division may not be sufficiently stable or abundant to persist and support cleavages after the first or several cell cycles. In this case, activating the genome by the second division cycle would circumvent this problem by producing the products required to sustain the subsequent cell division cycles.

PROMOTING EGA/ZGA BY TITRATION OF MATERNAL FACTORS OR NUCLEOCYTOPLASMIC RATIO IN *XENOPUS* AND ZEBRAFISH

Although the first cleavage is longer in *Xenopus* due to XWee1-dependent Map kinase (Mos) induced M phase delay (Murakami et al., 1999), the subsequent cleavages are rapid (i.e., cells divide every 30 minutes before zygotic genome activation). The time to accomplish the first division cycle is also longer in zebrafish embryos (i.e., about 45 minutes), while the subsequent cleavage cycles are rapid and occur every 15 minutes until zygotic genome activation (ZGA). Thus, the cells in fish and frog embryos can potentially undergo more cleavage cycles before degradation or depletion of a maternally supplied regulator would be expected. Indeed, a limiting maternally supplied cytoplasmic factor, or factors, and an increase in the DNA to cytoplasm ratio contribute to zygotic genome activation (Newport and Kirschner, 1982a).

A candidate transcriptional repressor that may become limiting due to increasing chromatin content is the methyltransferase *XDmnt1*. In support of Dnmt1 acting as a transcriptional repressor, Dmnt1 depletion in *Xenopus* is associated with premature zygotic transcription (Stancheva and Meehan, 2000). Repression of zygotic transcription in this context is independent of Dnmt1 methyl transferase activity as a mutant form of Dnmt1 lacking methylase activity suppresses the transcriptional repression defect (Dunican et al., 2008). In contrast to the clock mechanism operating in the mouse, and in support of chromatin associated depletion, introducing additional sperm, or polyspermy, also results in precocious genome activation (Newport and Kirschner, 1982b). In the mouse, Dnmt1 protects imprinted loci from global demethylation that accompanies genome reprogramming in early embryos and will be discussed later.

Maternal-effect mutants in zebrafish may help to clarify the contribution of each of these mechanisms. In zebrafish *acytokinesis* mutants, the nuclear division cycle lengthens even though cytokinesis is disrupted (Kishimoto et al., 2004). This suggests that zygotic genome activation can occur independent of cytokinesis because the transition from rapid synchronous cell cycles to lengthened mitotic cell cycles is usually associated with divisions that occur after zygotic genome activation. However, whether zygotic genes are activated as expected in mutants where cytokinesis is blocked will require experimental verification. In zebrafish *futile cycle* (*fue*) progeny discussed earlier, chromosomal segregation is compromised without blocking cytokinesis such that the nuclear: cytoplasm ratio is increased only within the few cells that contain DNA (Dekens et al., 2003). In these cells the zygotic genome is precociously activated. Thus, *futile cycle* mutants provide genetic support for nucleocytoplasmic ratio rather than maternal product depletion activating zygotic gene expression since zygotic genome occurs precociously at stages before the maternal product would be depleted in wild type (Dekens et al., 2003). Alternatively the maternal factor becomes depleted in the chromatin containing cells, but not in those cells lacking chromatin. It will be of interest to test whether Dmnt1 or another candidate regulator is differentially abundant in *fue* cells with DNA compared to those without.

Zebrafish maternal-effect *screeching halt* mutants arrest with compromised chromatin segregation and aberrant zygotic genome activation (Wagner et al., 2004). Abnormal chromatin is observed by the sphere stage during the midblastula transition MBT (i.e., after the ZGA and prior to gastrulation) in zebrafish (Wagner et al., 2004). Expression of a subset of the genes that should be activated at the ZGA indicates that *screeching halt* is not required for wholesale ZGA. However, the expressed genes show abnormal distribution. For example, *no tail/brachyury* is expressed, but is not restricted to the marginal blastomeres as it is in wild-type embryos (Wagner et al., 2004). Other genes, such as a ventral marker *tbx6*, fail to be expressed (Wagner et al., 2004). Expansion of *ntl* expression to nonmarginal cells and even to cells at the animal pole is intriguing. It is possible that abnormal expansion of *ntl* in *screeching halt* mutants reflects failure to activate the expression of a zygotic antagonist that would otherwise prevent *ntl* expression in cells distal to the margin (e.g., the Nodal antagonist Lefty, *the* homeobox protein Goosecoid, a Wnt antagonist, or an inhibitor for Fgf-mediated relay). Whether *screeching halt* mediates activation of a subset of zygotic genes or a more trivial explanation (e.g., activation of a checkpoint) underlies aberrant gene expression remains to be tested.

Regulation of gene expression at ZGA relies on proper cell division or nucleo: cytoplasm ratio in the zebrafish (Zamir et al., 1997); thus, the primary cell division defect in *screeching halt* mutants likely causes the dysregulated gene expression in *screeching halt* mutants. In addition to cell division and gene expression defects, *screeching halt* mutants fail to gastrulate (Wagner et al., 2004). This defect is also likely secondary to the cell division and gene expression defects (for example, lack of proper expression of an essential zygotic regulator of gastrulation). The identity of the disrupted gene should clarify the basis of the chromatin segregation and genome activation defects in zebrafish *screeching halt* mutants.

The zebrafish *janus* mutant is a spontaneous temperature-sensitive strict maternal mutation that disrupts a gene of unknown identity (Abdelilah et al., 1994). Janus regulates cohesion between blastomeres along the first cleavage plane as evidenced by mutant progeny with a shared yolk and two distinct half-blastoderms that do not mix with one another until epiboly (Abdelilah et al. 1994;

Abdelilah and Driever 1997). In light of the nucleo: cytoplasmic ratio and titration of a maternal factor by chromatin theories of zygotic genome activation, the *janus* embryos would be expected to activate zygotic genes at the same time as wild type. In this situation each half blastoderm would contain half as many cells as a wild-type blastoderm (i.e., the individual half-blastoderms of *janus* mutant progeny would be "younger" or have fewer cells per blastoderm at the time of ZGA, but the total number of cells per embryo would be the same as wild type). If the blastoderms of *janus* progeny are developmentally independent, the ZGA might be differentially activated in each half-blastoderm. Based on similar numbers of cells initially and the common yolk, the two blastoderms of *janus* mutant progeny are expected to function developmentally as one embryo in terms of cell cycle regulation and the onset of zygotic gene expression due to the shared yolk. However, when zygotic gene expression is initiated, it remains to be examined in *janus* mutant progeny.

Post-ZGA events, including epiboly and induction of mesendodermal and other tissue-specific markers in both half-blastoderms, indicate that Janus is not required to activate zygotic gene expression (Abdelilah and Driever, 1997; Abdelilah et al., 1994). Notably, the progeny of *janus* mutant females with twinned blastoderms that do not touch express germ ring markers around the entire circumference of the margin of each blastoderm. In contrast, those embryos with fused blastoderms maintain germ ring or margin marker expression only in regions of the outer margin devoid of contact between the twinned blastoderms (Abdelilah and Driever, 1997). It is possible that the sites of marginal contact between the blastodiscs are not interpreted as marginal. In this case, the T box gene *no tail/brachyury* is either not induced or is not maintained in this region of the embryo. Contact with neighboring cells might inhibit factors required to promote maintenance of the germ ring, or margin identity. Alternatively the regions in contact may not be accessible to a source of diffusible signal from the yolk. In *janus* embryos bifurcation of the blastoderm is random with respect to the dorsal organizer (i.e., embryos were observed with a dorsal organizer in one of the two or in both half blastoderms). Thus, studies of the *janus* mutant revealed that distinct mechanisms regulate dorsal–ventral axis formation and establish the cleavage planes of the embryo (Abdelilah et al., 1994). While it is clear that zygotic genome activation occurs, identification of the disrupted gene and further investigation of cell cycle control and the timing of ZGA in *janus* mutants are interesting questions worth pursuing.

• • • •

Reprogramming: Epigenetic Modifications and Zygotic Genome Activation

After fertilization of the mouse egg both the maternal and the paternal genomes are reprogrammed. In general demethylation of the genome correlates with permissive chromatin and pluripotency, and hemimethylation of the cytosine 5′ position of CpG dinucleotides correlates with repressive states or loss of pluripotency. The paternal genome is rapidly demethylated prior to the first cell division, while the maternal genome is protected and demethylation occurs in a stage-specific and stepwise manner in the mouse (Santos, Hendrich et al. 2002). Differential demethylation of the maternal and paternal genomes is regulated by maternal factors, which prevent demethylation of imprinted regions of the maternal genome. Imprinted genes tend to be clustered on the chromosome in so called imprinting control regions (ICRs) (Edwards and Ferguson-Smith, 2007).

Differential methylation of ICRs is a mechanism to promote parental-specific gene expression. These differential imprints (e.g., DNA methylation) are established in the germline (i.e., in oocytes and sperm) by methyltransferases (Figure 18A). Differential methylation must be maintained post-fertilization to permit normal embryonic development in mammals (Figure 18B). Maternal factors that regulate reprogramming and establishment of DNA methylation in oocytes and maintenance of differential methylation at ICRs in embryos will be discussed in this chapter.

DNMTs: DIFFERENTIAL METHYLATION AND GENE EXPRESSION

Multiple DNA methyltransferases, Dnmts, are present in mouse oocytes, Dnmt1 oocyte, Dnmt1o, an oocyte-specific Dnmt, Dnmt1s, Dnmt3a, and Dnmt3b (Howell et al., 2001; Hirasawa et al., 2008; Hirasawa and Sasaki 2009; Kaneda et al., 2004; Cirio et al., 2008; Cirio et al., 2008; Kurihara et al., 2008). Dnmt1o regulates maternal imprinting as evidenced by the reduced methylation of imprinted genes in the progeny of mutant mothers (Howell et al., 2001). However, the persistence of methylated imprinted genes in Dnmt1o progeny and the presence of multiple Dnmts in the ovary

A) Establishment of parental specific methylation profiles in germline

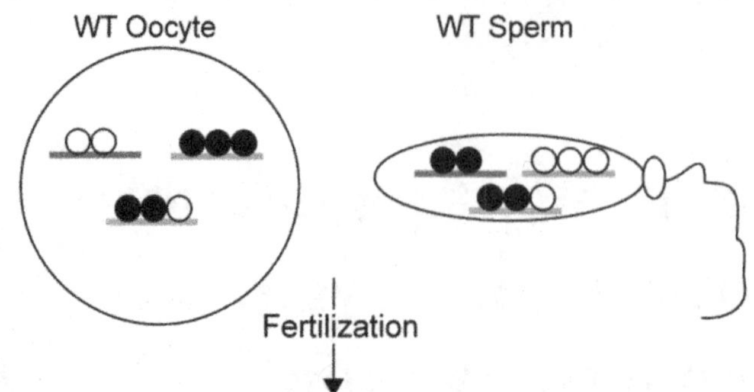

Loci specific requirements for Zfp57 to establish differential methylation of ICRs. Dnmt3a and Dnmt3b are required for differential methylation.

B) Maintenance of parental specific methylation profiles in germline

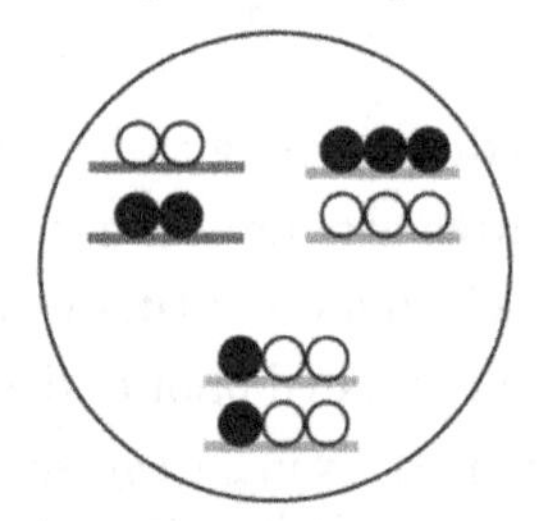

Global erasure of methylation, remethylation, and maintenance of differential methylation of ICRs requires MZ Zfp57, Dpp3a/Stella, and Dnmt1 and Dnmt1o.

C) MZ Zfp57 is required to establish and maintain differential methylation of ICRs

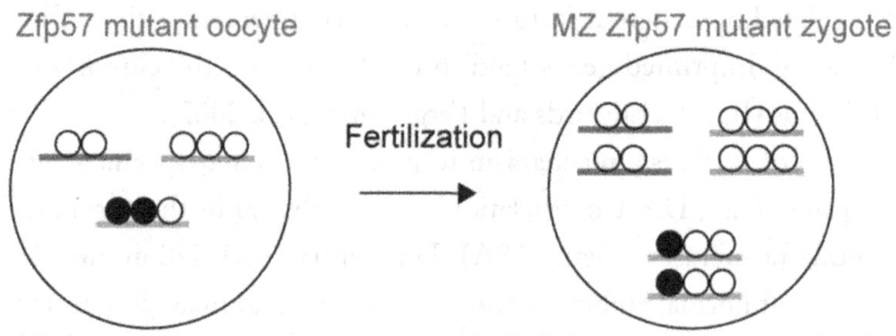

D) Maternal Dpp3a/Stella is required to maintain maternal imprints

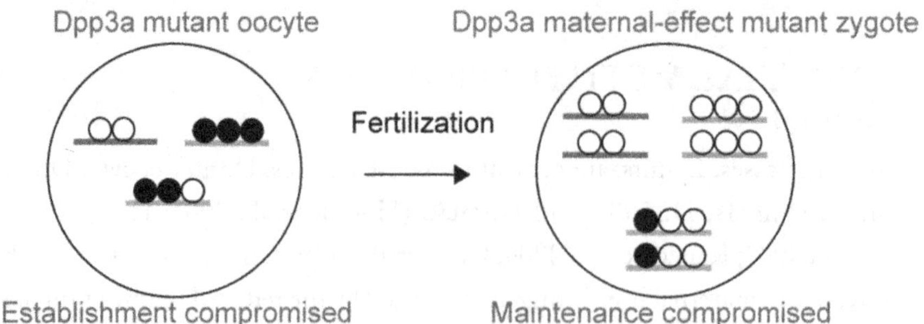

pointed toward roles for additional Dnmts in maintenance of imprinted loci in the mouse. Conditional knockout of both Dnmt3a and Dnmt3b with the oocyte-specific ZP3 promoter driving Cre completely eliminates maternal methylation during oogenesis (Hirasawa et al., 2008). In contrast, maternal methylation marks are maintained later in preimplantation stage embryos when the Dnmt3s are knocked out alone or simultaneously (Hirasawa et al., 2008). These findings showed that Dnmt3a and Dntm3b regulate establishment of maternal methylation in the germline, but other Dnmts function to maintain maternal imprints in the embryo.

The observation that most of the Dnmt1 protein present in preimplantation stage embryos is of maternal origin, and the similar partial loss of imprinted gene methylation between the Dnmt1 conditional maternal knockout and Dnmt1o mutants suggested that zygotic Dnmt1 might also contribute to maintenance of maternal methylation (Howell et al., 2001; Hirasawa et al., 2008). This notion was confirmed by the complete loss of differential methylation of imprinted loci when both maternal and zygotic Dnmt1 were eliminated (Hirasawa et al. 2008). Thus, both maternal and zygotic Dnmt1 contribute to maintenance of imprinting marks in preimplantation stage embryos. The mechanism by which Dnmt1 is recruited to differentially methylated regions, DMRs, and how DMRs are selectively protected during stepwise demethylation of the maternal genome in the midst of global demethylation is not understood.

Zfp57: ESTABLISHING MATERNAL MARKS AND MAINTAINING MARKS OF PARENTAL ORIGIN

Zfp57 encodes a Kruppel Associated Box (KRAB) Zinc finger protein. zfp57 is abundant in embryonic stem cells and is down regulated gene upon their differentiation (Li and Leder, 2007). Zfp57 function in DNA methylation at specific maternal loci within the germline is supported by direct binding of Zfp57 to the small nuclear ribonucleoprotein-associated protein N, snrpn, differential methylation region and reduced methylation at this DMR in maternal-zygotic mutants (Li et al., 2008). Normal methylation patterns of paternal imprints in the germline indicated that MZZfp57 is not required to establish paternal imprints. However, MZZfp57 maintains differential methylation

---

FIGURE 18: Schematic depicting establishment and Maintenance of Imprinted Loci. (A) Within the germline parental-specific DNA methylation profiles (filled circles = methylated; open spheres = unmethylated) are established at imprinted loci (blue and orange lines; green represents nonimprinted locus). (B) These imprints are inherited by the embryo. Reprogramming upon fertilization global demethylation and protection of maternally imprinted loci occurs. (C) Maternal Zfp57 is required to establish a subset of maternal imprints, and maternal and zygotic Zfp57 regulate maintenance of differential methylation at specific imprinting control regions, ICRs. (D) Maternal Dpp3a regulates maintenance of differential methylation in the zygote.

of both maternally and paternally imprinted loci as evident from the reduced expression of imprinted paternal genes and elevated expression of maternally imprinted genes in maternal-zygotic mutants (Figure 18C) (Li et al., 2008). Maternal Zfp57 is not strictly required to maintain differential methylation of maternally and paternally imprinted loci as zygotic Zp57 can suppress lethality; however, both maternal and zygotic contribution are required for optimal maintenance of methylation of imprinted regions after fertilization (Li et al., 2008). In other work, the KRAB domain when expressed with a co-repressor complex facilitated DNA methylation of a reporter transgene in early mouse embryos (Wiznerowicz et al., 2007). This de novo methylation activity together with the ability of zygotic Zfp57 to suppress the maternal defect in establishing imprints in the maternal germline indicates that maintaining methylation of ICRs may be an active process rather than a simple passive or protective mechanism.

Suppression of midgestation lethality and restored differential methylation by functional zygotic Zfp57 point toward additional mechanisms to mark the parental origin of the chromosomes to ensure parental-specific gene expression in the absence of loci-specific methylation in the maternal zfp57 mutants. Examples of "memory" of parental origin for imprinted loci even in the absence of or following erasure of DNA methylation marks implicate multiple epigenetic modifications in distinguishing a chromosomes parent of origin (Davis et al., 2000; Lucifero et al., 2004). For example, repressive histone marks have been connected to recognition of parental origin independent of DNA methylation (Lewis et al., 2004).

In humans, mutations in zfp57 are correlated with hypomethylation of ICRs and developmental defects including, diabetes, brain, and cardiovascular dysmorphology and function (Hirasawa and Feil, 2008; Mackay et al., 2008). The relatively late phenotypes in humans and early embryonic lethality in the mouse mutants are consistent with compromised development due to dysregulated gene expression as differential imprints are gradually lost.

Alternatively, heritable differential methylation of some loci may be required in a stage- or tissue-specific manner. This would account for late-stage as well as tissue-specific phenotypes as only those tissues that require parental-specific gene expression for normal development or function would be affected. In either case, the later phenotypes would likely be caused by inappropriate regulation of target genes as a consequence of lost imprinting.

DPPA3/STELLA/PGC7: PRESERVING MATERNAL MARKS

Dppa3, also known as STELLA and PGC7 is a putative DNA/RNA binding protein with a characterized nuclear localization sequence, NLS. The role of Dppa3 in preserving the DNA methylation of imprinted maternal loci during postfertilization epigenetic reprogramming was established by several independent studies of maternal-effect mutants; mutant progeny arrest due to abnormal cleavage prior to the two-cell stage (Nakamura et al., 2007; Bortvin et al., 2004; Payer et al., 2003).

Dppa3 shuttles between the nucleus and cytoplasm via its direct association with Ran binding protein 5 (Ranbp5) (Nakamura et al., 2007). Ranbp5-dependent shuttling of Dppa3 from the cytoplasm prior to fertilization to the male and female pronuclei after fertilization suggested that Dppa3 activity might be required in the fertilized embryo (Nakamura et al., 2007). However, the inability of zygotic Dppa3 to compensate for lack of the maternal protein is also consistent with a role prior to fertilization and genome activation. Manipulating the Ranbp5-dependent shuttling to retain Dppa3 in the nucleus, Nakamura and colleagues were able to suppress the dppa3 maternal-effect two-cell stage arrest (Nakamura et al., 2007). Development of the rescued progeny to blastocyst stages supports a role for Dppa3 after fertilization, but before the two-cell stage.

The mutant phenotypes provide evidence that a Dppa3-dependent mechanism acts after fertilization to protect the female pronuclear DNA, but not the male pronuclear DNA from demethylation (Figure 18D). Although the maternal pronucleus is not protected from demethylation comparable H3K9 methylation between Dppa3 mutants and wild type both before and after fertilization support H3K9-independent Dppa3 activity. Dppa3 likely acts in a locus-specific manner to preserve maternal methylation by a mechanism that remains to be determined as comparison of loci known to be imprinted, nonimprinted, repetitive, and nonrepetitive did not reveal a particular pattern or class of loci protected by Dppa3 (Nakamura et al. 2007).

Distinct maternal-effect genes (e.g., Dppa3/Stella, Zfp57) regulate establishment of specific maternally imprinted loci in the female germline, but act more broadly to maintain both maternal and paternal methylation profiles after fertilization. Based on the maternal-effect phenotypes discussed above it is clear that multiple genes and diverse epigenetic modifications contribute to establishment and maintenance of imprinted loci. The extent to which these distinct epigenetic modifications are redundant, are integrated or interpreted, and whether distinct tissues require different combinations of parental origin imprints for normal development remains to be determined.

MODIFYING HISTONE AND SWITCHING FROM MATERNAL TO ZYGOTIC HISTONES

Maternal products govern both activation and repression of the zygotic genome to prevent its precocious activation. One model to account for repression of the zygotic genome involves zygotic chromatin that is not competent to be transcribed due to maternally mediated Histone modifications. The repertoire of histone modifications consists of dynamic and reversible changes, including acetylation and phosphorylation as well as more stable modifications, such as methylation. Hype racetylation of Histones or methylation at lysine residues (H3K4, H3K36, H3K79) (Santos-Rosa et al., 2002; Krogan et al., 2003; Schubeler, MacAlpine et al., 2004) and methylation at Arginine residues (mono, di, or trimethylated) correlate with activation, while methylation at key lysine residues are repressive (H3K9, H3K27, H4K20) (Peters et al., 2003; Rice et al., 2003; Cao et al., 2002). These modifications of Histones occur individually or in combination with each other; thus, the combination of histone modifications simultaneously present at a given time produces diversity or a "histone code" (Jenuwein and Allis, 2001; Strahl and Allis, 2000; Turner, 2000). The transition from a highly differentiated oocyte to totipotent embryo is regulated by maternally supplied gene products that promote reprogramming (i.e., modulation of the gene expression potential or profile), and poise the embryo for totipotency, yet we only have hints as to how this transition is controlled in vertebrates.

Transcriptional repression in the mouse two-cell stage embryo has been linked to regulation of chromatin structure, which is partially mediated by changes in Histone H1 and the acetylation state of core Histones (Henery et al., 1995; Wiekowski et al., 1993; Wiekowski et al., 1997; Davis et al. 1996). Treating mouse embryos with drugs that promote hyperacetylation attenuates the repression of early-activated genes usually observed at the two-cell stage (Kijima et al. 1993; Kruh, 1982; Aoki et al., 1997). In addition, acetylation, which would act to overcome repression, is robust only at the two-cell stage in the mouse, when expression of zygotic genes has been activated (Aoki et al., 1997). The end of repression associated with the maternal pronucleus coincides with early ZGA (Wiekowski et al., 1997). Thus, maternal Histone is thought to mediate repression until the two-cell stage and later to cooperate with zygotic Histones to regulate repression after the one- to two-cell transition.

The embryos of Xenopus and zebrafish are also supplied with a large pool of maternal Histone. This maternal pool is proposed to be depleted by successive cell divisions, and to effectively repress transcription until zygotic genome activation at the mid-blastula transition (MBT) discussed earlier. In contrast to the situation in the mouse where zygotic genome activation is thought to be clock- dependent, zygotic genome activation in Xenopus and fish is coupled to cell divisions and nucleo: cytoplasmic ratio as exemplified by the zebrafish futile cycle mutants discussed earlier. In Xenopus, abundant maternal histones repress TATA-binding protein (TBP) activity (Veenstra et al., 1999) and reviewed by Nothias et al. (1995), while in zebrafish TATA-binding protein is only required to degrade a subset of maternal mRNAs present in the embryo (Ferg et al., 2007).

Demethylation of DNA and modification of chromatin state are associated with acquisition of pluripotency. Compared to mice, in embryos of zebrafish and Xenopus, the zygotic genome is activated later in development (i.e., relative to absolute cell number). At the time of ZGA the cells of the embryo are not committed to a particular fate and are therefore considered pluripotent. It is anticipated that Histone state may prevent zygotic gene expression before ZGA, and that modification of the Histone signature will be required for, or will accompany, acquisition of pluripotency at the time of ZGA. Genome wide chromatin immunoprecipitation, CHIP, analysis of Histone H3 in zebrafish and in Xenopus show that marks usually associated with activation (i.e., H3K4Me3), and repressive modifications (e.g., H3 lysine 27 trimethylation) were only detectable coincident with zygotic genome activation (Akkers et al., 2009; Vastenhouw et al., 2010). In the zebrafish study, bivalents, bearing both modifications and thought to be poised for activation, were also detected in embryos, but only after the midblastula transition or ZGA (Vastenhouw et al., 2010). Likewise RNA polymerase phosphorylation, an indicator active transcription, was only detectable in embryonic stages after zygotic genome activation (Vastenhouw et al., 2010). This fits well with the models proposed for maintenance of pluripotency in embryonic stem cells in which both modifications are observed simultaneously (Bernstein et al., 2006; Mikkelsen et al., 2007; Pan et al., 2007; Zhao et al., 2007). When both marks are present, the repressive marks are dominant, and the gene is not expressed. Such "poised" states require only the elimination of the repressive H3K27me3 mark to initiate gene expression after differentiation.

In contrast to the zebrafish embryo and ES cells, bivalents in the Xenopus study did not correlate with paused gene expression, leading to the conclusion that the detected bivalents are due to overlap in cell type-specific gene expression (i.e., a mixed population in which the gene is repressed in one cell type and actively expressed in another) (Akkers et al., 2009). Lack of detectable H3K27me3 until after the onset of gastrulation suggests that, in Xenopus, mechanisms other than poised bivalent promoters mediate the transition from pluripotent to differentiated states. Similar findings have also been reported in Drosophila. Specifically, the number of bivalents is only a minor population compared to promoters with individual marks (i.e., activating or repressive) (Schaner et al., 2003; Rudolph et al., 2007). Despite the differences between the findings of these studies, it is clear that the chromatin profile is dramatically modified at or coincident with zygotic genome activation. The mechanisms and proteins responsible for this zygotic chromatin signature remain to be determined, but it is expected that maternally supplied proteins will be among the regulators identified through functional and genetic studies.

SLBP: PRODUCING ENOUGH HISTONE

Stem loop binding protein, SLBP, is maternally expressed and is abundant in the oocytes of many species (Sanchez and Marzluff, 2004; Sullivan et al., 2001; Allard et al., 2002; Ingledue et al., 2000; Arnold et al., 2008; Zhang et al., 2009). In mouse oocytes, SLBP is detectable in meiosis II arrested oocytes through the first cell division of the fertilized embryo, after which its abundance is regulated in a cell cycle-dependent manner (Arnold et al., 2008; Zhang et al., 2009; Allard et al., 2002; Allard, Champigny et al. 2002). Stem loop binding protein regulates the stability and translation of Histone encoding RNAs by binding to the 3′UTR of histone messages to promote their stability and efficient translation (Whitfield et al., 2000; Wang et al., 1996). In the mouse, maternal SLBP ensures adequate amounts of histone are produced to promote DNA replication as evidenced from the failure of mutant mice to accumulate histones H3 and H4 and suppression of S-phase arrest by injecting of "whole" histone (Arnold et al., 2008). The block in S-phase was reported to occur without impacting histone 2A or histone 2B, suggesting some specificity (Arnold et al., 2008). Drosophila and C. elegans mutants deficient in SLBP arrest when the S-phase checkpoint is activated (Kodama et al., 2002; Lanzotti et al., 2002; Sullivan et al., 2001). Thus, it is possible that embryonic arrest of the mouse mutants is also due to activation of an S-phase checkpoint. Drosophila and C. elegans accomplish more cell cycles before the embryos succumb to chromosomal and spindle defects because the mechanism to detect these chromosomal defects does not operate until later developmental stages in these invertebrates.

Npm2: REGULATING NUCLEAR MORPHOLOGY AND UNMASKING mRNAS

Nucleoplasmin 2 (Npm2) was first identified in Xenopus as an oocyte-specific protein that binds histones and removes protamines, the arginine-rich proteins that replace histone in condensed sperm, to promote assembly of the nucleosome thus allowing replication of the paternal genome (Burns et al., 2003; Burglin et al., 1987; Dilworth et al., 1987; Dingwall et al., 1987; Gillespie and Blow, 2000; Ohsumi and Katagiri, 1991; Philpott et al., 1991). A role for Npm2 in promoting translation of maternal mRNAs in Xenopus has been proposed based on experiments in which histone H4 mRNA is not translated when expressed from a plasmid in oocytes and early embryos, but is translated if histone H4 mRNA is injected (Meric et al., 1997). The major protein associated with histone H4 containing ribonucleoprotein complexes was nucleoplasmin (Meric et al., 1997). Further biochemical analysis of Npm2 led the authors to propose a model whereby Npm2 promotes unmasking and thus translation of the masked mRNAs (Meric et al., 1997). Such a mechanism would be analogous to the role of hyperphosphorylated Npmn2 in unpacking sperm chromatin after fertilization.

In the mouse maternal Npm2 regulates nucleolar formation, chromatin condensation, and exit from the first mitotic division (Burns et al., 2003). Aberrant nuclear morphology and chromatin condensation defects are evident in mutant oocytes and can be seen in the diffuse localization of the nucleolar protein fibrillarin and diminished hypoacetylated histone H3, an indicator of compact chromatin at this stage (Burns et al., 2003). Burns and colleagues demonstrated that transcription-requiring complex components were present, which is consistent with reports of others that blocking transcription or histone deacetylases does not prevent the first mitotic division in the mouse (Ma et al. 2001). The survival of a few progeny to birth in the absence of maternal Npm2 by a mechanism that is independent of zygotic Npm2 hints at compensation by other oocyte expressed Npms or epigenetic factors. The extent to which the oocyte defect causes the embryonic phenotype or correlates with survival to birth is not clear. Components of the pathway and mechanism by which maternal Npm2 regulates early development in the mouse including whether Npm2 promotes translation as predicted by the unmasking model in Xenopus remain to be discovered.

HR6A: A ROLE FOR DNA REPAIR IN DEVELOPMENT BEYOND THE ZGA

HR6 is a Rad-related ubiquitin conjugating DNA repair enzyme. HR6 is a component of the post-replication repair machinery, also known as the replication damage bypass, because it facilitates escape of cells from S-phase arrest by permitting replication across lesions. In the mouse and in humans there are two HR6 genes, HR6A and HR6B, which have redundant zygotic functions during embryonic development, but have independent contributions with regard to fertility (Roest, van Klaveren et al. 1996; Roest, Baarends et al., 2004; Koken, Smit et al., 1992). Both are broadly expressed, but HR6A is more abundant in oocytes, and only HR6A is maternally required for development beyond the two-cell stage in mouse (Roest et al., 2004). Mutant females are not compromised in oocyte production, ovulation, or fertilization (Roest et al., 2004). H3K9 methylation of the maternal pronucleus, and H3K4 methylation of one-cell stage embryos does not require HR6A; thus, developmental arrest is independent of changes of the H3 methylation profile (Roest et al., 2004). It is possible the two-cell block is due to defective transcription or regulation of another chromatin modification (e.g., ubiquitination). However, the mechanism by which maternal HR6A regulates development beyond the two-cell stage remains unclear and awaits further analysis of the maternal phenotype.

SWI/SNF/TIFα: CHROMATIN REMODELING TO ACTIVATE ZYGOTIC GENE EXPRESSION

Maternal TIFα, transcription intermediary factor, translocates from the cytoplasm to the pronucleus by the two-cell stage, coincident with the first phase of zygotic genome activation. TIFα is required for the promoter localization of RNA polymerase II and two members of the SWI (mating type switching)/SNF (sucrose nonfermentation) family of ATP-dependent chromatin remodelers, the helicase SNF2H (sucrose nonfermentation 2) and BRG1 (brahma-related gene 1), an ATP-dependent helicase also known as SMARCA4. Evidence that SNF2H, BRG1, and TIFα promote transcriptional activation of a subset of genes comes from impaired transcriptional activation phenotypes of mutants disrupting any component individually (Torres-Padilla and Zernicka-Goetz, 2006; Bultman et al., 2000; Bultman et al., 2006). Diminished transcription is largely independent of global H3H4 or H3K9 acetylation in SNF2H and TIFα mutants (Torres-Padilla and Zernicka-Goetz, 2006; Bultman et al., 2000). BRG1 disruption is associated with reduced H3K4 (Bultman et al., 2006), which may reflect a more general role for BRG1 compared to the other complex components. The transcriptionally compromised genes between SNF2 and TIFα depleted embryos overlap; SNF2 impairs transcription of a subset of the genes affected by TIF depletion and biochemical evidence supports a mechanism whereby SNF2 mediates TIFα activated transcription of the overlapping genes (Torres-Padilla and Zernicka-Goetz, 2006). The TIFα cofactor or partner required to activate the remaining SNF2-independent genes has not been identified thus far. Cumulatively, the effects on zygotic gene activation independent of global Histone modification and the greater number of targets impacted when TIFα is depleted support a model whereby SWI/SNF complexes function downstream of Histone acetyltransferases (HATs) and TIFα to promote gene expression during zygotic genome activation.

The presence of TIFα-independent gene expression suggests that TIFα is not the master regulator responsible for wholesale zygotic gene expression. If a master regulator indeed exists its identity has not been described thus far. Given the distinct phases of EGA in the mouse it is not surprising that there are multiple regulators, or at least multiple cofactors associated with regulators of EGA. Such a network of regulators would ensure that genes are activated in the proper temporal order.

The dual bromodomain and WD repeat containing protein, Brwd1, also known as Wdr9 in mouse, interacts with Brg1, and was identified as a regulator of fertility in males and females through a forward genetic screen in the mouse (Huang et al., 2003; Lessard et al., 2004; Philipps et al., 2008). Mutant females develop advanced stage oocytes; however, these oocytes are partially blocked in meiosis II (Philipps et al., 2008). When inseminated, the progeny of brdw1 mutant females fail to develop to the blastocyst stage. Based on its association with Brg1 and the transcriptional activator, function of its polyglutamine rich domain maternal Brwd1 may promote developmental progression by activating transcription through its association with Brg1 and the SWI/SNF complex (Huang et al., 2003). Alternatively, Brwd1 could regulate chromatin remodeling and condensation to promote global transcriptional silencing accompanying maturation via its bromodomains. The mechanism by which Brwd promotes developmental competence remains to be determined.

ROLES FOR ZINC FINGER PROTEINS IN PROMOTING GENE EXPRESSION

Zygotic arrest 1 (Zar1) protein consists of a zinc-finger-like plant homeodomain (PHD) domain, suggesting a potential function as a regulator of transcription or modulator of chromatin. In mice maternal Zar1 regulates pronuclear fusion as evidenced from the failed pronuclear fusion in fertilized eggs of zar1 mutant females (Wu et al., 2003; Wu et al., 2003). A small fraction of eggs do successfully complete fertilization but show diminished levels of the Transcription-requiring complex (TRC). The TRC is an assay surveying a group of proteins (via S-35-Met labeling) whose synthesis depends on embryonic transcription, thus the TRC is used to provide an indication of zygotic genome activation (Wu et al., 2003). Whether diminished TRC abundance reveals a second function of Zar1 in activation of the zygotic genome, or this defect is a secondary consequence of an earlier phenotype is unclear. An argument that lends support for a potential second function for Zar1 is the observation that sperm contribution is dispensable for ZGA in mice (i.e., zygotic genome activation occurs even in parthenogenetically activated eggs). Until Zar1 interacting proteins and targets are identified the mechanism and pathway through which Zar1 regulates pronuclear fusion and whether Zar1 has additional functions in ZGA remain unclear.

CTCF is a zinc finger DNA binding protein known to act as a transcriptional activator, repressor, and epigenetic regulator of imprinting. As a vertebrate insulator protein, CTCF binds to CTCF binding sites found at the imprinting control region of maternally imprinted genes to regulate access to enhancers of the maternal and paternal chromosomes (Fedoriw et al., 2004; Bell et al., 1999; Pant et al., 2003; Szabo et al., 2004; Yoon et al., 2005; Hikichi et al., 2003; Fitzpatrick et al., 2007; Chao et al., 2002). When bound by CTCF methylation of the maternal imprinting control region, ICR, is blocked. This favors expression of the maternal allele over the paternal allele. In oocytes transgenic RNAi lines implicate CTCF in promoting polar body elimination during meiosis I and meiosis II. Later, in embryos CTCF promotes zygotic gene expression as indicated by impaired TRC abundance and microarray evidence that genes enriched for CTCF binding sites are down regulated in CTCF depleted oocytes (Wan, Pan et al. 2008). CTCF interaction with the large subunit of RNA polymerase II, and co localization with the Pol II complex in the nucleus provide support for a mechanism whereby CTCF promotes gene expression by recruiting Pol II to maternal ICRs to promote gene expression (Wan et al., 2008; Chernukhin et al., 2007). Pronuclear transfer experiments and RNA rescue at the one-cell stage provided evidence that defective chromatin organization or epigenetic changes are either reversible or do not contribute to the maternal-effect phenotype. Specifically, CTCF depleted nuclei can support development in a wild-type cytoplasm, and conversely wild-type nuclei are unable to rescue the CTCF depletion phenotype (Wan et al., 2008). From these experiments the authors infer that transcriptional changes rather than chromatin modifications are likely responsible for delayed development at the two-cell stage (Wan et al., 2008). In the future, it will be important to determine which transcriptional changes are significant and directly due to depletion of CTCF.

Multiple mechanisms contribute to control of gene expression. This includes repression of zygotic genes until embryonic genome activation. In mammals where imprinting occurs this includes epigenetic modifications to promote the differential expression of maternal and paternal genes. Epigenetic modifications (DNA methylation, histone methylation) together with small RNA regulators of gene expression provide multiple levels of regulation. However, we are only beginning to understand the individual and relative contributions of each of these maternally provided regulatory mechanisms and their developmental impacts.

• • • •

Dorsal–Ventral Axis Formation before Zygotic Genome Activation in Zebrafish and Frogs

While the animal–vegetal axis is patterned exclusively by maternal factors, the later developing dorsal–ventral axis depends on formation of the animal–vegetal axis and requires both maternal and zygotic contributions in zebrafish and frogs. In contrast, in the mouse dorsal–ventral axis formation is regulated by zygotic factors only. As discussed earlier in the chapter on oocyte polarity, many of the proteins and mRNAs known to localize along the animal–vegetal axis of oocytes or eggs of *Xenopus* and zebrafish are regulators of the dorsal–ventral axis of the embryo (Figure 19). Oocyte depletion studies in *Xenopus* and maternal-effect and maternal zygotic zebrafish mutants have revealed crucial roles for these localized maternal products during embryogenesis for proper development of the germ layers, and the head to tail (anterior–posterior), back to belly (dorsal–ventral), and left-right embryonic axes. These genes are summarized in Table 2 and will be discussed in the following chapters.

THE ANIMAL–VEGETAL OOCYTE AXIS: POSITIONING AND TRANSLOCATION OF DORSAL DETERMINANTS

Before the eggs of frogs and fish are fertilized the dorsal determinants are already positioned asymmetrically at the vegetal pole. The dorsal determinants remain at the vegetal pole until after fertilization when they undergo translocation to the prospective dorsal side (Figure 14). The precise details of the mechanisms of translocation are distinct, but many key elements are shared. Microtubule arrays are assembled in early zebrafish and *Xenopus* embryos. When embryos are treated with inhibitors that disrupt microtubules soon after fertilization the consequence is defective establishment of the dorsal–ventral axis (Elinson and Rowning, 1988; Jesuthasan and Strahle, 1997; Rowning et al., 1997). Similarly, exposing early embryos to UV irradiation perturbs the microtubule cytoskeleton, and causes ventralized phenotypes in the embryos of fish and frogs. Ventralized phenotypes are characterized by diminished expression of dorsal organizer markers and concomitant

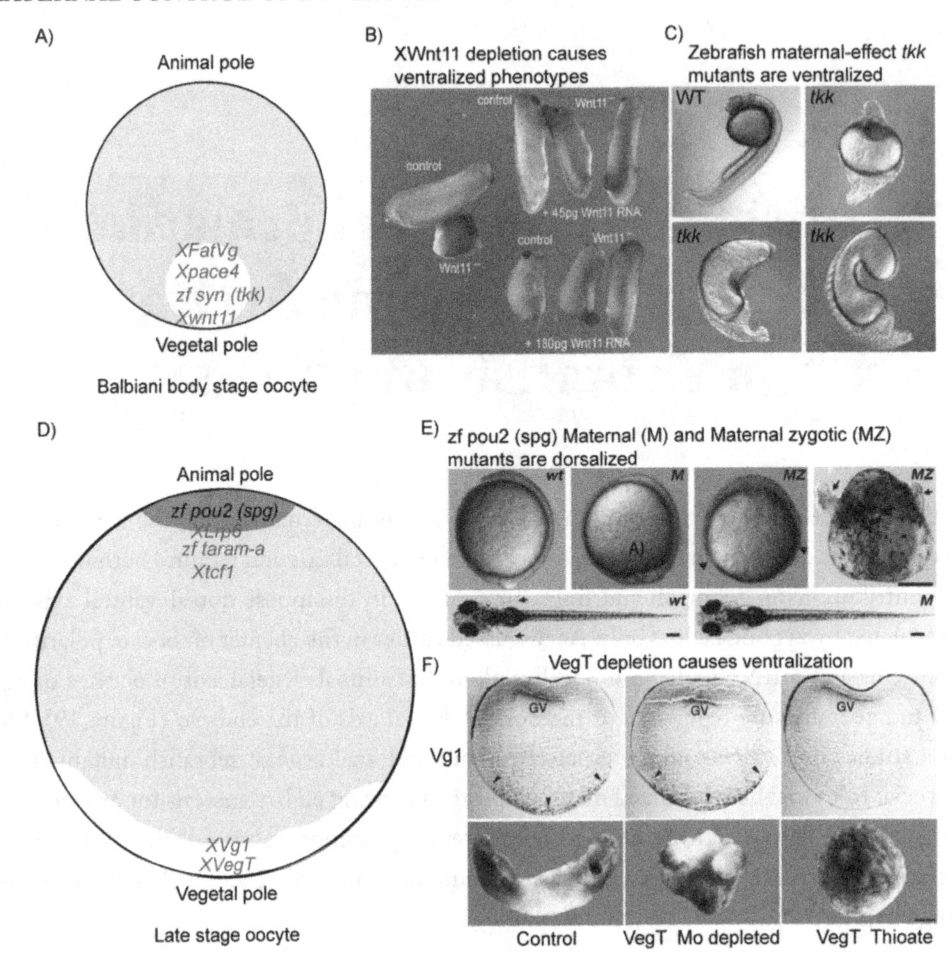

FIGURE 19: mRNAs that localize asymmetrically along the animal–vegetal axis in oocytes and their essential roles in embryonic patterning and axis formation. (A) mRNAs that localize to the Balbiani body of early oocytes contribute to dorsal (blue) formation. (B) Maternal depletion of Wnt11 in *Xenopus* or (D) maternal-effect *tkk* mutants disrupt *syntabulin* and cause ventralized phenotypes of embryos. mRNAs that localize after Balbiani body dispersal contribute to dorsal (blue) and ventral (black) axis formation in *Xenopus* and zebrafish. Panel B is from Tao et al. (2005b), panel C is from Nojima et al. (2004), panel E is from Lunde et al. (2004), panel F is from Heasman et al. (2001).

expansion of ventral markers in early embryos. Later when morphologically distinct structures are evident, ventralized embryos show expansion of ventral derived structures (e.g., blood, fin, tail structures) at the expense of dorsal derived structures (e.g., axial mesoderm and head structures; Figure 18). The early cytoskeletal perturbations established a role for microtubules as components of the dorsal determination translocation machinery. Recent depletion studies in *Xenopus* oocytes and zebrafish maternal-effect mutants have begun to trace genes whose products have essential roles in dorsal–ventral axis formation back to the oocyte.

Cytoplasmic transplantation studies in *Xenopus* identified the vegetal pole as the source of the dorsal inducing activity or determinants (Fujisue et al., 1993; Holowacz and Elinson, 1993; Marikawa et al., 1997). Similarly elegant embryo recombination experiments demonstrated the ability of grafted yolk to induce mesoderm in adjacent host blastoderm tissue in zebrafish (Mizuno et al., 1996). In related studies, this mesoderm inducing capability was localized to the yolk positioned closest to the vegetal pole within the first 20 minutes after fertilization (Mizuno et al., 1999). When these recombination experiments were performed at later developmental stages, mesoderm-inducing activity no longer resided in the vegetal region. As discussed in the previous chapter, zygotic genome activation occurs at a much later developmental stage in zebrafish and in *Xenopus*; thus, the source of the mesoderm-inducing activity has long been appreciated to be a maternally supplied or regulated factor. In *Xenopus* and zebrafish, the initial localization of the dorsal determining activity at the vegetal pole relies on the animal–vegetal pattern established earlier during oogenesis; however, the identity of the dorsal determinants and the molecular pathways and mechanisms mediating their localization are not fully understood.

MATERNAL WNTS AND THE DORSAL–VENTRAL EMBRYONIC AXIS-SIGNALING

Wnt family ligands are generally classified into two or three groups (i.e., canonical, noncanonical, and calcium). The canonical Wnts signal through Frizzled receptors to promote stabilization and nuclear accumulation of β-Catenin. The noncanonical Wnts are sometimes divided into two groups, the Planar Cell Polarity, PCP, Wnts and Calcium Wnts. Noncanonical Wnts signal through Frizzled receptors to activate JNK or Calcium to modulate cell morphology or behavior via the cytoskeleton, and to orient structures or position cells within a field (For more details on zygotic functions of PCP Wnts the reader is referred to recent reviews (Fanto and McNeill, 2004; Barrow, 2006; Chien et al., 2009; Mlodzik, 2002; Wallingford, 2005; Roszko et al., 2009; Tada and Kai, 2009; Jenny and Mlodzik, 2006; Kohn and Moon, 2005). Canonical Wnt or β-catenin activity is sufficient to induce the dorsal organizer in over expression studies in *Xenopus* (Cui et al., 1995; Ku and Melton, 1993; Wylie et al., 1996; McMahon and Moon 1989; Guger and Gumbiner, 1995; Jenny, 2010; Sokol et al., 1991). Those Wnts that show axis inducing activity and stabilize β-Catenin in gain of function studies are designated as canonical, while those that do not impact patterning or β-Catenin stability are considered to be noncanonical. However, the situation is likely more complicated as some Wnt ligands seem to impact both β-Catenin mediated patterning and PCP activities, the two pathways share downstream components, and canonical and noncanonical Wnts may work together to mediate developmental processes. This chapter focuses on the maternal roles identified for Wnt pathway components during dorsal–ventral axis patterning in *Xenopus* and zebrafish embryos.

ICHABOD/β-CATENIN TRANSCRIPTIONAL COMPLEX

Although it would be more "canonical" to begin with the ligands, we will start with β-catenin as it has a long history as an indicator of dorsal fate. To date, a complete understanding of dorsal determination remains elusive; however, evidence from studies of fish and frogs demonstrate that Canonical Wnt signaling pathway components, Dishevelled (Dsh) in particular, promotes stabilization of β- catenin (β-cat) in the nuclei of cells that will develop as dorsal (Funayama, 1995; Rowning et al., 1997; Schneider et al., 1996; Miller et al., 1999). Studies in which β-catenin was depleted in *Xenopus*, and in zebrafish maternal-effect mutants disrupting β-*catenin2*, *ichabod*, or that compromise factors acting upstream of β-catenin stabilization on the prospective dorsal side (e.g., *hecate* [discussed further below], *tokkaebe*, *brom bones*/*hnRNPI* mutants) evinced the essential role of β-*catenin2* in dorsal–ventral axis formation (Heasman et al., 1994; Wylie et al., 1996; Hedgepeth et al., 1997; Kofron et al., 2001; Wagner et al., 2004; Lyman, Gingerich et al., 2005; Tao et al., 2005; Kofro et al., 2007; Mei et al., 2009; Kelly et al., 1998; Nojima et al., 2004; Bellipanni et al., 2006). Conversely stabilizing β-Catenin with Lithium chloride induces ectopic axes (Stachel et al., 1993; Klein and Melton, 1996; Larabell et al., 1997), and reviewed by Solnica-Krezel (1999) and Hibi et al. (2002). In the dorsal nuclei, a conserved β-catenin transcriptional complex composed of β-catenin, pygopus, and Bcl9, also known as *legless*, promotes the expression of genes required to specify axial mesoderm (e.g., nodal target genes, antagonists of ventral fates to be discussed later) (Belenkaya et al., 2002; Carrera et al., 2008). Thus, the blastula organizer requires and is molecularly defined by the nuclear accumulation of maternal β-catenin in zebrafish and frog embryos.

As mentioned, *bozozok*/ *dharma* is a zygotic gene whose transcripts are detected at stages prior to embryonic genome activation in zebrafish. *Bozozok* is a transcriptional target of β-*catenin* encoding a homeodomain transcriptional repressor that is essential for dorsal organizer formation (Ryu et al., 2001; Leung et al., 2003). Bozozok/Dharma was identified as a binding partner for a maternal E3 ubiquitin ligase protein, ligand of numb protein-X (Lnx-1), in yeast two-hybrid and pull down assays (Ro and Dawid, 2009). Evidence that Lnx-1 ubiquitinates and contributes to selective degradation of Bozozok by proteolysis, includes stabilization of Bozozok/Dharma and dorsalization of embryos depleted of maternal *lnx-1* using translation blocking morpholinos (Ro and Dawid 2009). Attenuation of Lnx-1 dorsalization phenotypes in Lnx-1/Boz double morphants, and suppression of Boz gain of function phenotypes when *lnx-1* is co-expressed with *boz* provide additional support for a role of Lnx-1 as a post-translational regulator of Boz stability in zebrafish embryos (Figure 20) (Ro and Dawid, 2009).

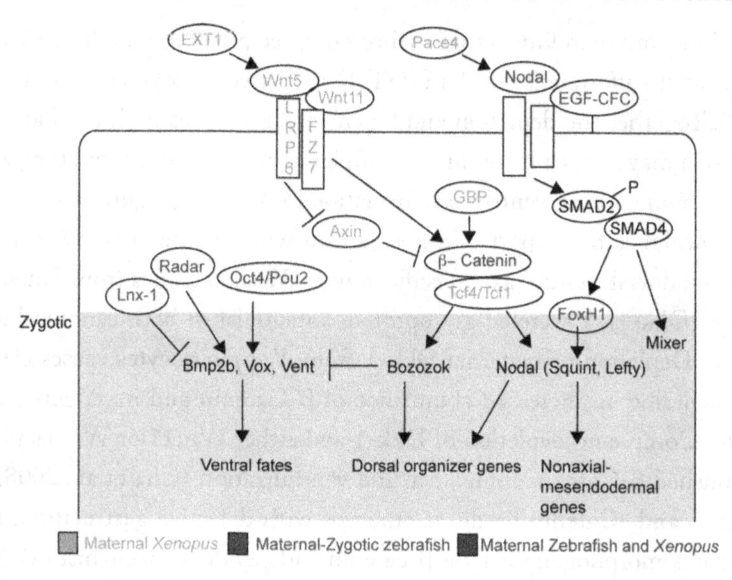

FIGURE 20: Schematic of maternal pathways regulating dorsal–ventral pattern in *Xenopus* and zebrafish.

Wnt LIGANDS

For many years, the maternal Wnt ligand acting upstream of Dsh and β-Catenin during dorsal specification was not known. Several Wnt ligands are maternally provided to the embryos of *Xenopus* and zebrafish so functional studies are required to identify the key Wnt ligand(s) mediating dorsal–ventral axis formation (Cui et al., 1995; Krauss et al., 1992; Ku and Melton, 1993; Makita et al., 1998; Moon, 1993; Rauch et al., 1997). Among the maternally expressed *wnts* in *Xenopus*, *Wnt11* localizes to the oocyte vegetal pole via the early Balbiani body mediated pathway (Figure 19) (Ku and Melton, 1993). Later in embryos, Wnt11 is enriched on the dorsal side as a consequence of cortical rotation (Tao et al., 2005). Wnt11, when ectopically expressed in embryos, is sufficient to promote dorsalization through a mechanism that is dependent on maternal β-catenin (Tao et al., 2005). Conversely, depletion of maternal Wnt11 from late stage *Xenopus* oocytes leads to axis defects in embryos, and the ventralized phenotypes are more profound when both maternal and zygotic Wnt signaling are depleted (Figure 19) (Tao et al., 2005). Thus, maternal and zygotic Wnt11 contribute to dorsal–ventral patterning of *Xenopus* embryos.

Wnt5a is also expressed in *Xenopus* oocytes; however, its RNA is not asymmetrically distributed until gastrulation stages (Moon, 1993). Although originally identified as a noncanonical Wnt based on its gain of function phenotype (Moon, 1993), maternal depletion studies in *Xenopus* reveal a role for Wnt5a in β-Catenin stabilization and dorsal–ventral axis formation (Cha et al., 2008). Wnt5a is not necessary for *wnt11* localization in oocytes suggesting that it does not promote dorsal specification by positioning *wnt11* mRNA (Cha et al., 2008). Biochemical analysis indicates that

Wnt11 and Wnt5a form homodimers that co-immunoprecipitate as a multimeric complex dependent on tyrosylprotein sulfotransferase-1 (TPST-1) mediated tyrosyl sulfation (Cha et al., 2008; Cha et al., 2009). Together the depletion and biochemical analyses indicate that maternal Wnt11 and maternal Wnt5a may act as a heteromeric complex to mediate axis formation, as either alone is not sufficient to promote dorsal–ventral axis formation in *Xenopus* (Figure 20).

Additional evidence in support of a mechanism whereby maternal Wnt11 and Wnt5a act together to promote dorsal–ventral axis specification in *Xenopus* comes from functional analysis of maternal Dikkopf (Dkk-1), a secreted glycoprotein antagonist of both canonical and noncanonical Wnt signaling. Depletion of maternal Dkk-1 from *Xenopus* oocytes causes elevated canonical Wnt signaling, including an increased abundance of β-Catenin and its targets (e.g., *xnr3*, *xnr5*) (Cha et al., 2008). Compound depletion of Dkk-1 and either Wnt11 or Wnt5a all cause the same phenotype, diminished β-Catenin abundance and ventralization (Cha et al., 2008). Simultaneous depletion of Dkk-1 and -Catenin produces embryos with defective gastrulation; thus, indicating that Dkk-1 regulates morphogenesis by a β-catenin-independent mechanism (Cha et al., 2008). The targets of Dkk-1 that mediate β-catenin-independent functions during early *Xenopus* development are not known.

While these depletion studies show a clear need for Wnt11 and Wnt5 supplied to the oocyte, *wnt11*, but not *wnt5*, transcripts are asymmetrically localized. *Wnt11* mRNA first localizes to the Balbiani body, then later it localizes to the vegetal pole of *Xenopus* oocytes (Ku and Melton, 1993). Thus, an interesting question is whether Wnt11 also has a necessary function in the oocyte. Alternatively, it could be localized to the Balbiani body of early oocytes to protect the transcript from degradation or to position Wnt11 for its later role in dorsal–ventral patterning of the embryo. However, addressing any role of Wnt11 in early-stage oocytes will require depletion prior to or at stages when the Balbiani body is present, which remains technically difficult.

Due to partial duplication of the genome, zebrafish have two *wnt11* genes, one of which is disrupted in *silberblick*, *slb*, mutants and is necessary for gastrulation movements (Heisenberg et al., 2000). Loss of maternal and zygotic Slb/Wnt11 does not cause dorsal–ventral patterning defects; thus, Slb/Wnt11 is not essential for axis development. The second gene, *wnt11r*, is not expressed during early cleavage stages, so it too is unlikely to be the functional equivalent of *Xenopus* Wnt11.

Another Wnt ligand, Pipetail/Wnt5, is present as a maternally provided message in zebrafish (Kilian et al., 2003). In one study, zebrafish mutants lacking maternal and zygotic Wnt5/Ppt showed defective axis formation; however, the embryos were dorsalized rather than ventralized (Westfall et al., 2003). A dorsalized phenotype is more consistent with Wnt5 functioning as an antagonist of dorsal specification rather than acting as a, or to induce a dorsal determinant. Wnt5 has been proposed to antagonize the dorsal promoting β-catenin pathway by modulating the calcium concentration or competing for receptors; however, the mechanism and targets mediating antagonism are not fully understood. In another study utilizing a presumed null mutant allele that produces an early truncation of Wnt5, loss of maternal and zygotic Wnt5 did not cause dorsal–ventral patterning defects (Ciruna et al., 2006). Moreover, compound mutants lacking maternal and zygotic Ppt/Wnt5, Slb/Wnt11, and also depleted of Wnt4 did not cause dorsal–ventral patterning phenotypes (Ciruna et al., 2006). It is possible that the different phenotypes can be attributed to the nature of the mutant alleles used in each study. For example, the allele in the Westfall study may be a neomorphic allele or produce a stable truncated protein that interferes with the Wnt/β-Catenin pathway. Additional biochemical studies and creating a maternal zygotic mutant with the two alleles in trans could clarify the situation.

The maternal expression of Wnt ligands and Frizzled receptors, together with gain of function and interference strategies indicate a potential role of maternal Wnt-Fz-mediated signaling in dorsal–ventral specification in zebrafish. However, to date an essential maternal Wnt ligand and receptor have yet to be identified for dorsal–ventral axis formation in zebrafish.

RECEPTORS AND CO-RECEPTORS OF MATERNAL Wnts

Evidence that XFrizzled7 (XFz7) may the receptor for maternal Wnt11 during axis formation in *Xenopus* includes biochemical interaction between XFz7 and Wnt11, potentiation of Wnt11 induced dorsalization when co-expressed with XFz7 in gain of function studies in *Xenopus* embryos, oocyte XFz7 antisense depletion, and morpholino phenotypes (i.e., loss of dorsal mesoderm and anterior structures) (Djiane et al., 2000; Sumanas et al., 2000; Sumanas and Ekker, 2001). Canonical Wnt signaling requires Low-density lipoprotein receptor-related protein (LRP) co-receptors. Two LRP receptors, LRP5 and LRP6, are maternally supplied *Xenopus*; however, only antisense depletion of maternal LRP6 causes elevated Axin protein abundance, failure to accumulate β-Catenin, and ventralized phenotypes that resemble maternal Wnt11 depletion (Houston and Wylie, 2002; Kofron et al., 2007).

Consistent with roles as receptor and essential co-factor, excess Wnt11 is unable to suppress the LRP6 depletion phenotype, while β-Catenin overexpression attenuates the ventralized phenotypes caused by depletion of Fz7 or LRP6 (Sumanas et al., 2000; Kofron et al., 2007). Whether other maternally expressed Fz receptors are involved awaits functional investigation. In zebrafish, several Fz receptors are maternally supplied, but data to assess their essential maternal function are not available.

The glycosyl transferase Exostin, XExt1, is a maternally expressed copolymerase enzyme that adds glycosaminoglycans, GAGs, essential for herparin sulphate proteoglycan function and activation of Wnt11 (Tao et al., 2005). A function for XExt1 in maternal Wnt signaling is supported by the similar depletion phenotypes of maternal XExt1 to those of Wnt11 and β-catenin depletion, and the enhanced ventralized phenotypes observed when Wnt11 and XExt1 are partially co-depleted (Tao et al., 2005). Suppression of XExt1 and Wnt11 phenotypes by β-catenin established a role for extracellular Wnt signaling to promote β-Catenin accumulation and dorsal fate specification in *Xenopus*, while the maternal Wnt and potentiating factors in zebrafish remain to be discovered.

CYTOPLASMIC FACTORS REGULATING MATERNAL β-CATENIN

Maternal Axin is a component of a complex that targets β-catenin for degradation through phosphorylation. Depletion phenotypes of maternal Axin are opposite to phenotypes due to depleting maternal Wnt11, and identify Axin as a negative regulator of maternal β-Catenin during dorsal–ventral patterning in *Xenopus*. When maternal Axin is depleted, β-catenin is stabilized and its nuclear accumulation expands to ventral regions, Wnt target gene expression is elevated, and, consequently, the embryos are dorsalized (Willert et al., 1999; Kofron et al., 2001; Nusse, 2005). These dorsalized phenotypes can be partially rescued when Axin is provided back to ventral but not to dorsal cells (Kofron et al., 2001).

Evidence that maternal Wnt11 and LRP6 act in the oocyte to regulate β-Catenin abundance comes from the phenotypes in depleted oocytes, and the finding that LRP6 and Wnt11 depletion phenotypes can only be suppressed when LRP6 is supplied prior to fertilization. In Axin, depleted oocytes β-catenin and Wnt targets are elevated suggesting that LRP6 promotes dorsal development at least in part by regulating Axin in oocytes (Kofron et al., 2001; Kofron et al., 2007). Simultaneous depletion of LRP6 and Axin attenuates elevated Wnt target expression relative to Axin depletion alone (Kofron et al., 2007). This finding is consistent with a role for LRP6 in limiting Axin availability; however, whether the mechanism involves sequestration, degradation of Axin protein, or both remains to be fully resolved.

GBP, XGSK-3 binding protein, is a cytoplasmic inhibitor of glycogen synthase kinase-3, GSK-3. GBP binds to GSK-3 and prevents it from phosphorylating its substrates (Yost et al., 1998). Antisense depletion of maternal GBP causes ventralized phenotypes, while excess GBP leads to β-catenin stabilization, ectopic dorsal axis induction, and restores axis formation in UV irradiated embryos (Yost et al., 1998). Local action of XGSK-3 binding protein has been proposed to protect β-catenin from degradation induced by the XGSK-3/Axin complex, thus facilitating β-catenin stabilization specifically in the dorsal nuclei. Asymmetric accumulation of β-catenin has not been reported in oocytes or prior to cleavage stages in embryos. This would suggest a mechanism whereby GBP is only translated during cleavage stages, and specifically in the future dorsal cells. Alternatively, GBP is post-translationally repressed or is only activated in a localized manner. How maternal XGSK-3 binding protein is regionally regulated is not understood; however, evidence that GBP binds to kinesin light chain (Klc) and translocates toward dorsal during cortical rotation includes co-immunoprecipitation of tagged GBP and Klc proteins, and the movement of GBP–GFP fusion protein particles toward dorsal (Weaver et al., 2003). These findings support model whereby GBP is inactive until it reaches the prospective dorsal side, and then once activated GBP would alleviate GSK3 repression of β-catenin thus allowing its nuclear accumulation.

Together the maternal depletion phenotypes of these positive and negative regulators of β-catenin abundance indicate that β-catenin is regulated in the oocyte for normal dorsal–ventral axis patterning in *Xenopus*. This is surprising given the observations that β-Catenin accumulation occurs in cleavage stage embryos, but not earlier, and that β-Catenin can only rescue Wnt11 or LRP6 depletion when supplied after fertilization (Kofron et al., 2007). Moreover, depleting β-catenin during cleavage stages or blocking its translation in the vegetal pole is sufficient to cause ventralization of *Xenopus* embryos. When considered together these findings indicate that the pattern established in the oocyte must not be fixed, or is not sufficient for proper dorsal specification without *de novo* production of β-catenin protein.

A mechanism whereby Wnt11, LRP6, and Axin act in the oocyte to localize mRNAs encoding activators or repressors of β-catenin along the oocyte axis would prepare the embryo to locally produce β-catenin in one of the vegetal cells at a later developmental stage. Such a mechanism would account for regulation of β-catenin in the oocyte, and for the observation that β-catenin can rescue ventralization when supplied back to the embryo after fertilization. Controlling β-catenin abundance in the oocyte might also provide robustness to the dorsal specification program. Future studies are necessary to determine if other components of the Wnt signaling pathway are also localized along the animal–vegetal axis of oocytes, and also contribute dorsal–ventral axis formation in *Xenopus* embryos.

INTRACELLULAR TRANSDUCERS OF Wnt SIGNALS

The T-cell factor (Tcf)/lymphoid enhancing factor (Lef) family of transcription factors are key transducers of Wnt signaling. TCF proteins are able to interact both with co-repressors (Groucho) or co-activators (β-catenin). In *Xenopus* three Tcf/Lef family members (XTcf1, XTcf3, and XTcf4) are maternally expressed and contribute to dorsal–ventral axis patterning as activators or repressors of Wnt signaling (Standley et al., 2006; Houston et al., 2002).

Maternal XTcf4 functions as an activator of Wnt target gene expression. Simultaneous partial depletion of the two maternal XTcf4 isoforms diminishes the expression of Wnt target genes, and causes ventralized phenotypes (Standley et al., 2006). Importantly, these hypomorphic depletion phenotypes can be attenuated by injecting XTcf4 mRNA at stages prior to fertilization, demonstrating loss of dorsal structures is due to diminished XTcf4 (Standley et al., 2006).

In contrast to XTcf4, XTcf1 and XTcf3 act as repressors of Wnt target genes in ventral regions. Depletion of either XTcf1 or XTcf3 results in elevated expression of the organizer genes, Siamois and Xnr3 (Houston, Kofron et al. 2002; Standley et al., 2006). Expansion of dorsal markers is β-catenin and VegT-dependent manner as simultaneous depletion of maternal β-catenin and XTcf3 or maternal β-catenin both result in ventralized phenotypes (Houston et al., 2002; Standley et al., 2006). Similar phenotypes between XTcf1 and XTcf3 single and compound depleted oocytes argue for cooperative repressor activity rather than independent contributions (Standley et al., 2006).

In contrast to XTcf3, XTcf1 has an additional requirement to promote Wnt target expression in dorsal tissue (Standley et al., 2006; Houston et al., 2002). Thus, XTcf1 acts as a repressor on the ventral side, and conversely functions as an activator of Wnt targets on the dorsal side. These findings illustrate that multiple members of the Tcf transcription family can have divergent activities, acting as repressors, activators, or as both a repressor and an activator, to exert opposing effects on a process (e.g., dorsal–ventral patterning in *Xenopus*) in a cell type- and context-dependent manner. However, the extent to which differences in activity reflect the binding affinity of each Tcf or regional variation in cofactor availability will require domain swapping, biochemical analysis, or identification of the key maternal cofactors through genetic or maternal depletion approaches.

Several maternally expressed *tcf* genes have also been identified in zebrafish including *lef1*, *tcf7*, and *tcf3*, but thus far a maternal function has only been described for *tcf3* (Veien et al., 2005; Dorsky et al., 1999; Kim et al., 2000). Homozygous mutants for a *tcf3* loss of function allele called *headless* have relatively weak zygotic phenotypes (e.g., normal body axis but smaller eyes) and are semi-viable. Thus, some mutants reach adulthood, and the maternal function of *tcf3* was assessed (Kim et al., 2000). Loss of maternal and zygotic *tcf3* in zebrafish causes expansion of *wnt8* and the absence of head structures (Kim et al., 2000). Depletion of maternal Tcf3 in *Xenopus* does not prevent head formation possibly because zygotic Tcf3 remained intact in the *Xenopus* depletion studies. As discussed earlier, maternal β-catenin signaling is crucial for specification of the zygotic dorsal organizer during early development of *Xenopus* and zebrafish embryos. In gastrula stage zebrafish embryos, *wnt8* expression is specifically excluded from the organizer proper, but persists in the lateral margin (Kim et al., 2000). At this stage, canonical Wnts, like Wnt8, have potent posteriorizing activity; excess Wnt due to overexpression or loss of inhibitors causes loss of anterior structures (Dorsky et al., 2003; Erter et al., 2001; Fekany-Lee et al., 2000; Heisenberg et al., 2001; Kelly et al., 1995; Kim et al., 2000; Lekven et al., 2001). In zebrafish maternal and zygotic Tcf3 repress Wnt activity and prevent posteriorization of the neurectoderm to allow development of anterior structures, such as the head.

MATERNAL WntS AND THE DORSAL EMBRYONIC AXIS-TRANSLOCATION MACHINERY

Movement of the dorsal determinants from vegetal to dorsal where β-catenin is stabilized occurs shortly after activation and requires microtubules as discussed earlier (Figure 14). Thus, one expects regulators of microtubule assembly and stability, motor proteins, and other microtubule binding proteins to be among the maternal gene products required to position the dorsal determinants and to mediate or direct their dorsal-ward movement. The mutants and depletion studies discussed below disrupt components of the translocation machinery, and consequently cause defective patterning of the dorsal–ventral axis.

HECATE: A ZEBRAFISH MATERNAL-EFFECT GENE REQUIRED FOR NUCLEAR ACCUMULATION OF β-CATENIN

Evidence that Hecate promotes dorsal–ventral axis formation in zebrafish comes from the ventralized maternal-effect phenotypes of the progeny of *hecate* mutant females (Lyman Gingerich et al., 2005). Defective nuclear accumulation of β-catenin, failure to activate the expression of dorsal organizer genes, and rescue of *hecate* ventralized phenotypes by components of the Wnt/β-catenin signaling pathway identify hecate as an upstream component of the dorsal determinant localization machinery or as a novel component of the Wnt pathway (Lyman Gingerich et al., 2005). The molecular identity of the disrupted gene in *hecate* mutants; however, has not been reported.

TOKKAEBE/SYNTABULIN: A ROLE IN DORSAL DETERMINANT POSITION AND RELEASE

In zebrafish Tokkaebi/Syntabulin is a motor linker protein required for translocation of the dorsal determinants to the dorsal side where they promote nuclear accumulation of β-catenin and expression dorsal target genes (Nojima et al., 2010; 2004). Syntabulin protein is localized to the vegetal pole of fertilized eggs and subsequently shifts from a vegetal position to a more lateral vegetal position in a microtubule-dependent manner (Nojima et al., 2010). Dynamic localization of syntabulin protein, and the ventralized phenotypes of *syntabulin* mutants identified a new step in dorsal determinant translocation (Figure 14). The initial vegetal pole localization of syntabulin protein and its molecular identity as a motor linker protein, and biochemical evidence demonstrating syntabulin's association with the Kif5b kinesin motor protein suggest that Syntabulin anchors the still elusive dorsal determinants to the vegetal region (Figure 14) (Nojima et al., 2010). As the egg is activated and the vegetal microtubule array is assembled, an asymmetric shift in syntabulin localization was observed and suggests that syntabulin mediates asymmetric displacement of the dorsal determinants within the vegetal pole toward the future dorsal side (Figure 14) (Nojima et al., 2010). Degradation of syntabulin protein after attaining an asymmetric position within the vegetal pole, suggests a model whereby Syntabulin protein degradation releases the dorsal determinants to traverse along the microtubule cytoskeleton toward the blastoderm. When the dorsal determinants arrive at the blastoderm margin, they activate β-catenin and the pathways necessary to specify and pattern dorsal (Figures 14 and 19) (Nojima et al., 2010).

In zebrafish *syntabulin* RNA is localized to the Balbiani body of primary oocytes and later to the vegetal pole of eggs by a mechanism that depends on Bucky ball mediated animal–vegetal polarity (Nojima et al., 2010). Notably, although *syntabulin* is localized to the Balbiani body where germ plasm also localizes in early zebrafish oocytes, *syntabulin* is not a component of germ plasm in embryos (i.e., it does not localize to the cleavage furrows where germ plasm localizes in early embryos, and later it is not localized to the primordial germ cells), and does not seem to be required for germ cell formation. If the existing *syntabulin* mutant allele is a protein null, then Syntabulin protein is also not required for polarity of the oocyte. Instead *syntabulin* seems to be localized in the oocyte specifically for its later role in patterning the dorsal–ventral embryonic axes.

KINESIN: MOVING TOWARD VEGETAL

In zebrafish, Kif5b is expressed in oocytes and binds to syntabulin; however, whether Kif5b functions to localize syntabulin in oocytes or is only required later for dorsal determinant localization in embryos awaits a maternal mutant disrupting zebrafish Kif5b. During later stages of oogenesis in *Xenopus* and zebrafish, a late vegetal pole transport pathway carries patterning molecules to the vegetal pole (recently reviewed by King et al., 2005; Kloc and Etkin, 2005; Minakhina and Steward, 2005). Experiments conducted in oocytes of *Xenopus*, in which kinesin is disrupted with antibod-

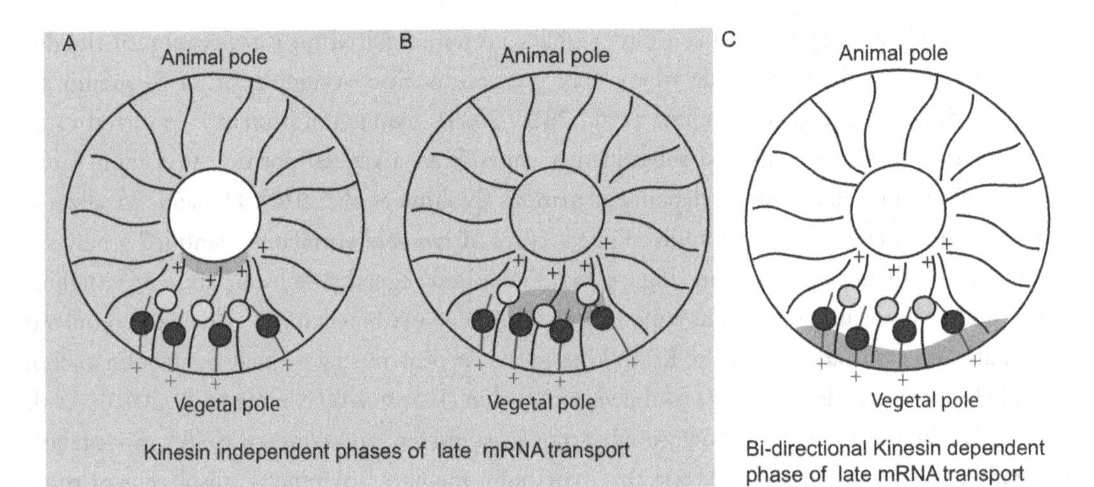

FIGURE 21: Vegetal transport of late localizing vegetal pole mRNAs involves (A, B) kinesin-independent and (C) kinesin-dependent phases. (A) Initially late localizing vegetal mRNAs are in a perinuclear position. (B) Movement of Vg1 from the nucleus to the midpoint between the nucleus and the vegetal cortex occurs independent of kinesin. (C) Movement from the midpoint to the vegetal cortex is bidirectional (red and black microtubules are of opposite orientation as indicated) and distinct kinesin motors (yellow and blue circles) contribute to Vg1 movement and docking at the cortex.

ies to block the motor domain of kinesin-2 or kinesin-1, or with mutant forms of a kinesin-related motor protein, Eg5, disrupt localization of *Vg1* RNA to the vegetal pole via the late vegetal pole transport pathway (Betley et al., 2004; Messitt et al., 2008). Through these studies Messitt and colleagues discovered that the initial steps of vegetal transport occur independent of kinesin function; however, both kinesin-1 and kinesin-2 are required to mediate localization during a previously unknown step in the vegetal localization pathway. This step regulates transport from the midpoint between the nucleus and vegetal cortex to the vegetal cortex (Figure 21). The presence of anti-parallel microtubules in the vegetal hemisphere and essential roles for two plus-ended motors indicates that transport could be bidirectional from the midpoint to the vegetal cortex; thus, docking would be key to enriching mRNAs at the vegetal cortex (Figure 21) (Messitt et al., 2008). Although the consequences of disrupting kinesin motor proteins to axial patterning were not reported in these studies, based on defective Vg1 localization, kinesin interference would be expected to cause axis specification defects similar to Vg1 depletion.

PROPER EGG ACTIVATION IS REQUIRED FOR DORSAL DETERMINANT LOCALIZATION

As discussed above, the dorsal determinants of fish and frogs are positioned at the vegetal pole of the oocyte during oogenesis where they remain through ovulation and fertilization. During egg activation, the determinants are moved to the future dorsal side by a microtubule-dependent mechanism (Figure 14). The temporal relationship between egg activation and dorsal determinant localization indicate that egg activation could contribute to effective dorsal determinant translocation. Genetic evidence from zebrafish maternal-effect mutants, like *hecate* mutants, with impaired dorsal determinant translocation have normal egg activation (e.g., cortical granule exocytosis, chorion elevation). However, zebrafish *brom bones* mutants discussed earlier for the necessary role of hnRNPI in egg activation and exocytosis of cortical granules provides an example of impaired egg activation disrupting dorsal determinant translocation. Compromised cortical granule exocytosis in *brom bones* progeny results in persisting cortical granules (Mei et al., 2009). The continued presence of the cortical granules interrupts the microtubule network and effectively impedes translocation of the dorsal determinants from the vegetal cortex to the prospective dorsal side; this result s in defective accumulation of nuclear β-catenin and consequently ventralization (Mei et al., 2009). The extent or severity of the maternal-effect ventralized phenotypes correlates with the strength of the earlier egg activation defect (i.e., the stronger the egg activation defect the stronger the degree of ventralization) (Mei et al., 2009).

The degree of the microtubule-dependent dorsal-ward shift of cytoplasm during cortical rotation of *Xenopus* eggs was previously shown to anticipate the size of the later forming Spemann–Mangold (dorsal) organizer (Figure 22) (Gerhart et al., 1989). Recently, oocyte depletion studies have identified two maternal factors required for cortical rotation and dorsal determinant translocation by mechanisms acting upstream of or in parallel to Wnt11 signaling. Oocyte depletion of FatVg, a lipid vesicle associated protein also known as Adipophilin, disrupts cortical rotation and causes abnormal accumulation of β-catenin at the vegetal pole, which leads to ventralization (Figure 22) (Chan et al., 2007). Based on its depletion phenotype and identity as a vesicle-associated protein FatVg is proposed to regulate vesicular trafficking concomitant with egg activation. Accordingly FatVg would mediate delivery of vesicles containing the molecular cargo required to stabilize β-catenin and promote its nuclear accumulation. Consistent with a role for FatVg in trafficking a component required to stabilize β-catenin, supplying excess β-catenin is not sufficient to rescue FatVg depletion (Chan et al., 2007).

Similar to FatVg diminution, depleting maternal *trim36*, a tripartite motif (composed of three domains, a ring finger domain a B-box zinc finger, and a coiled coil domain) containing ubiquitin ligase disrupts cortical rotation and ultimately results in failure to specify the dorsal organizer

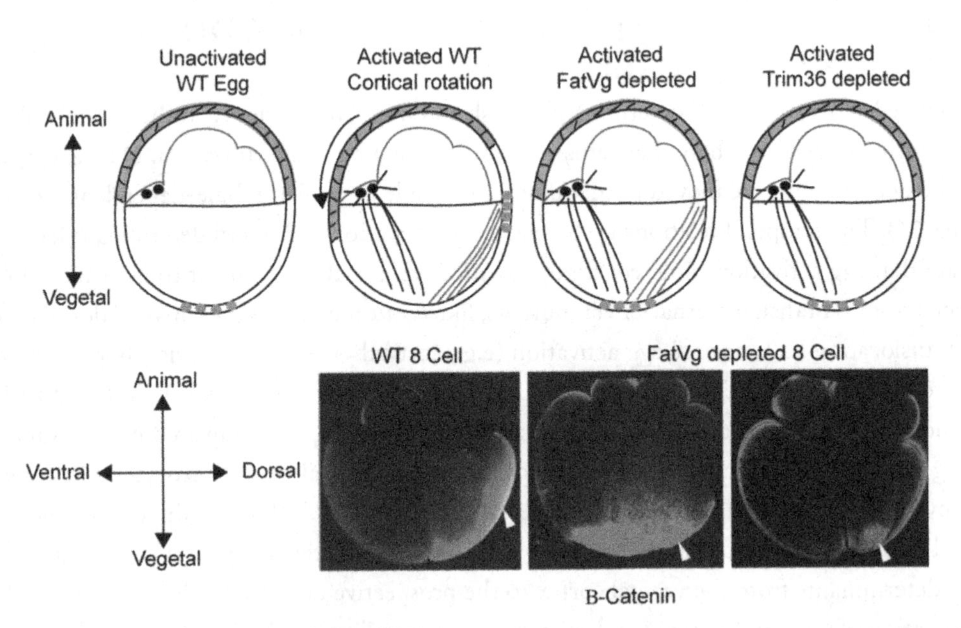

FIGURE 22: Schematics depicting the position of the dorsal determinants at the vegetal pole prior to cortical rotation and their microtubule-dependent movement during cortical rotation. FatVg-depleted oocytes are defective in cortical rotation and consequently dorsal determinant translocation is blocked. Trim36 is required to organize the vegetal microtubule array, and when depleted, the consequence is failed cortical rotation and dorsal determinant translocation. Images from Chan et al. (2007) show dorsal biased beta-catenin accumulation in wild-type *Xenopus* embryos, and inappropriate beta-catenin at the vegetal pole or reduced accumulation in embryos that were depleted of maternal FatVg.

causing ventralized phenotypes (Cuykendall and Houston, 2009). The diminished vegetal microtubule array in *trim36*-depleted oocytes and its restoration by wild type but not an ubiquitin ligase defective Trim36 provided evidence that Trim36 regulates polymerization of the vegetal microtubule required for cortical rotation and dorsal determinant translocation (Figure 22) (Cuykendall and Houston 2009). Phenotypic suppression studies place Trim36 upstream of Wnt11-mediated β-catenin activation; however, Trim36 is not sufficient to promote Wnt11/β-catenin signaling in gain of function studies (Cuykendall and Houston 2009). Thus, Trim36 may act in parallel to Wnt11, or more likely acts by a mechanism dependent on its ubiquitin ligase activity to promote microtubule organization because its ubiquitin ligase activity is required to rescue the axis defects.

Notably, both FatVg and Trim 36 localize to the Balbiani body of oocytes. Given their respective roles in localizing vesicles and organizing the vegetal microtubules, it will be interesting to learn whether FatVg and Trim36 have similar functions in oocytes, or are positioned in anticipation of their later roles in cortical rotation. Similar to observations in zebrafish *brom bones* mutants, which impede dorsal determinant localization as a consequence of impaired egg activation, FatVg-depleted embryos show a correlation between the severity of the cortical rotation defect and the strength of the ventralization phenotype. Thus in fish and frog failure to fully activate the egg compromises dorsal–ventral patterning by impeding translocation of determinants.

• • • •

Maternal Tgf-β and the Dorsal–Ventral Embryonic Axis

In addition to maternal Wnt signaling, patterning of the dorsal–ventral and anterior–posterior axes in *Xenopus* and zebrafish is accomplished through the contribution of TGF signaling pathways (Figure 20). Members of the transforming growth factor-β (TGFβ) signaling family, including Nodal and Bone morphogenetic proteins (BMPs), their transmembrane receptors, and the downstream transducers of these pathways show patterning defects when their zygotic gene function is lost (Schier and Talbot, 2005; White and Heasman, 2008). In some cases, these phenotypes are exacerbated by the additional loss of maternal contribution in zebrafish and *Xenopus* (Table 2; reviewed by Schier and Talbot, 2005; White and Heasman, 2008). In this section the emphasis is on the maternally regulated aspects and contributions of these pathways to axis formation.

TGFβ/BMP/NODAL: THE LIGANDS AND COFACTORS

Mutants disrupting Nodal pathway components indicate a predominantly zygotic role for Nodal signaling during dorsal–ventral patterning in zebrafish. Maternal function of the ligand *squint/nodal related 2* (*sqt/nr2*), and EGF-CFC proteins, essential extracellular membrane GPI tethered cofactor proteins for TGFβ signaling molecules (e.g., Nodal, Vg1, growth differentiation factor 1, GDF1, lefty/antivin nodal antagonists) are not strictly required for axis formation in zebrafish (Zhang et al., 1998; Schier and Shen, 2000; Yeo and Whitman, 2001; Yan et al., 2002; Cheng et al., 2003; Tanegashima et al., 2004).

Complete loss of Nodal signaling by eliminating the maternal and zygotic function of the EGF-CFC essential cofactor, One-eyed pinhead, results in more severe phenotypes (i.e., absence of head and trunk mesoderm and endoderm) compared to the zygotic mutant phenotypes (lack endoderm, prechordal mesoderm and floor plate) (Gritsman et al., 1999). Thus, maternal One-eyed pinhead facilitates normal development because the presence of maternal product ameliorates the phenotype caused by loss of zygotic function. In contrast to *MZoep* phenotypes, which are exacerbated by loss of maternal product, the phenotypes of mutants lacking both maternal and zygotic *sqt/nr2* are comparable to those caused by zygotic loss of function alone (Pei et al., 2007). Morpholino depletion studies are in agreement with the genetic data that translation of maternal *sqt*/*nr2* mRNA is not required for dorsal specification; however, depletion of *sqt*/*nr2* from unfertilized eggs has also been reported to cause more severe phenotypes compared to *MZsqt* mutants (Gritsman et al., 1999; Gore et al., 2005; Bennett et al., 2007; Hagos et al., 2007). Other studies indicate a role for maternal *squint*/*nr2*, but only in certain genetic backgrounds and environmental conditions (e.g., genetic backgrounds that compromise canonical and noncanonical Wnt signaling, elevated temperature, reduced Heat shock protein 90a) (Pei et al., 2007; Pei and Feldman, 2009; Sirotkin et al., 2000a). However, outside of these particular genetic and environmental contexts, the genetic data do not support a strict maternal requirement for *sqt*/*nr2* or *cyclops* (*cyc*)/*nodal related 1* (*nr1*) when zygotic Nodal signaling is intact. Embryos lacking maternal *sqt* or maternal *cyc* or both are viable and do not show patterning defects (Bennett et al., 2007; Hagos et al. 2007). Thus, maternal nodal contribution in zebrafish likely ensures sufficient nodal abundance to allow proper development of the dorsal organizer for formation of head and trunk mesoderm and endoderm, and to position the later developing anterior–posterior axis of the embryo.

In *Xenopus*, Lefty has been reported to act as a Wnt inhibitor and to bind to the EGF-CFC protein, known as FGF receptor ligand 1 (FRL1) (Tanegashima et al., 2004; Tao et al., 2005; Branford and Yost, 2002). *Xenopus* embryos depleted of maternal FRL1 are ventralized (Tao et al., 2005). These phenotypes resemble depletion of maternal Wnt11 or β-catenin and can be rescued by β-Catenin microinjection (Tao et al., 2005). FRL1 gain of function does not disrupt patterning, but potentiates dorsalization caused by gain of Wnt11 in *Xenopus* (Tao et al., 2005). Physical interaction between the proteins, and simultaneous depletion of maternal *Wnt11* and *FRL1* suggest that in *Xenopus* FRL1 has additional functions as a cofactor for Wnt/β-catenin signaling during dorsal axis formation.

Vg1 is one of the earliest examples of a localized RNA in vertebrate oocytes and eggs (Melton, 1987; Melton et al., 1989; Tannahill and Melton, 1989; Yisraeli and Melton, 1988; Yisraeli et al., 1989). Vg1, also known as growth differentiation factor 1, Gdf1, is a TGF-β family protein maternally supplied to the early embryo in its precursor form (Helde and Grunwald, 1993; Tannahill and Melton, 1989). Mature Vg1 is generated by site-specific cleavage to produce the active form; thus, processing Vg1 in a spatially and temporally specific manner would limit its activity and restrict mesoderm formation within the embryo. Depletion of maternal Vg1 from *Xenopus* oocytes using an antisense oligo approach impairs gastrulation and specification of axial mesoderm without perturbing the maternal Wnt pathway (Birsoy et al., 2006). Vg1 also promotes mesoderm induction and anterior development independent of BMP signaling (Birsoy et al., 2006). Vg1 instead is implicated in signaling through a Nodal pathway, where it promotes phosphorylation and activation of Smad2, which together with FoxH1 regulates the zygotic expression of genes required for anterior development after ZGA in fish and mouse (Pogoda et al., 2000; Sirotkin et al., 2000b; Hoodless et al., 2001;

Yamamoto et al., 2001). Zebrafish Vg1/DVR-1, and Vg1 RBP (RNA binding protein) are also maternally supplied and are localized in oocytes (Helde and Grunwald, 1993; Zhang et al., 1999). Mutants disrupting maternal Vg1/DVR-1 are not currently available in zebrafish; however, treating embryos with drugs that block signaling through Alk4, 5, and 7 receptors indicates that Activin like signals, including Vg1/DVR-1 only function after activation of the zygotic genome in zebrafish (Hagos et al. 2007).

XPace4: PROCESSING AND ACTIVATING MATERNAL TGFβ

Maternal XPace4, a pro-protein convertase also known as subtilisin-kexin like pro-protein convertase, Spc4, cleaves TGF proteins, and is localized to the Balbiani body of early *Xenopus* oocytes and later to the vegetal pole (Birsoy et al., 2005). In *Xenopus*, maternal XPace4 promotes mesoderm induction, blastopore closure, and development of anterior structures by processing TGFβ ligands as is evident from compromised processing of HA-tagged versions of Xnr1, Xnr2, Xnr3, and Vg1 in XPace4 depleted oocytes (Birsoy et al., 2005). In the same assays, XPace4 was dispensable for processing HA-tagged Xnr5, ActivinB, or Derrière thus indicating some degree of specificity in ligand processing by XPace4 (Birsoy et al., 2005). Zebrafish *pace4* is also expressed in the ovary, but whether it is localized in oocytes or contributes to axis patterning is not yet known. Maternal knockout of Pace4 in the mouse has not been reported. The zygotic loss of function phenotype in mouse is highly variable; however, genetic interactions with Nodal mutants indicate that zygotic Pace4 contributes to nodal-dependent anterior–posterior axis development (Constam and Robertson, 2000).

THE TRANSCRIPTION FACTORS

Zic2 is an abundant maternal zinc finger transcription factor related to the Odd-Paired class, first identified based on their segmentation phenotypes in *Drosophila* (Nusslein-Volhard and Wieschaus, 1980). Maternal Zic2 represses or limits mesoderm formation to allow proper patterning of the embryonic axis as is apparent from the delayed blastopore closure, exogastrulation (e.g., when involution of dorsal mesoderm and vertical interactions with the ectoderm are disrupted, the result is external chordamesoderm connected to neurectoderm), and complex neural patterning defects involving loss of forebrain that manifest in depleted oocytes (Houston and Wylie, 2005). These exogastrulation and forebrain defects are caused by excess Nodal signaling (primarily *Xnr5/6*), and can be attenuated by the Nodal antagonist Cerberus-short, which specifically binds Nodal, or by simultaneous depletion of VegT (Houston and Wylie, 2005; Piccolo et al., 1999). How Zic2 mediates VegT-dependent regulation of Nodal signaling is not understood, but it is likely indirect as Zic2 does not repress VegT in animal cap assays (Houston and Wylie, 2005). It remains possible that Zic2 regulates VegT through a target that is not present in animal caps, or Zic2 could promote a negative feedback loop to attenuate Nodal signaling after Nodal has been initiated by VegT. Lefty2 and Antivin are known to contribute to negative feedback regulation of Nodals during gastrulation in mouse, frogs, and fish; however, the abundance of these antagonists was not examined in embryos depleted of maternal Zic2 (Cheng et al., 2000; Meno et al., 1999).

FoxH1, also known as Fasth1, is a forkhead binding and SMAD binding domain containing transcriptional cofactor for SMAD proteins that acts to transduce Nodal signaling (Pogoda et al., 2000; Sirotkin et al., 2000b; Yamamoto et al., 2001; Hoodless et al., 2001). Fox H1 is maternally supplied in *Xenopus* and in zebrafish (Kofron et al., 2004; Pogoda et al., 2000). Depletion of maternal FoxH1 from *Xenopus* oocytes ablates head structures and axial mesoderm; a phenotype that is similar to maternal depletion of Vg1 (Kofron et al., 2004). Zebrafish mutants disrupting maternal *foxH1*, also known as *schmalspur*, are phenotypically normal; thus, in zebrafish embryos maternal *foxH1* is not strictly essential when paternally supplied or zygotic expression of *foxH1* is intact (Pogoda et al., 2000; Sirotkin et al., 2000b). However, maternal-zygotic *foxH1* zebrafish mutants show more severe phenotypes, including diminished organizer gene expression, a phenotype not observed in zygotic mutants, and are unable to respond to β-catenin gain of function (Pogoda et al. 2000; Sirotkin, Gates et al., 2000). Initial induction of β-catenin targets including Nodal and Nodal target expression occurs even in *foxH1* mutants; however, FoxH1 is required to maintain both the positive and negative feedback loops induced by Nodal signaling (Pogoda et al., 2000).

The maternal-zygotic phenotypes of zebrafish mutants disrupting *foxH1* are less severe when compared to the maternal depletion phenotypes in *Xenopus*, and the zygotic phenotypes of mouse *foxH1* mutants vary from mild (similar to *schmalspur*) to severe (resembling *MZoep*). The comparatively mild phenotype of the zebrafish *MZfoxH1* phenotypes, and the nature of the mutant allele raise the possibility that FoxH1 might mediate processes via functional domains outside of the DNA binding domain; these domains are predicted to be intact, even in MZ*sur/foxH1* mutants (Pogoda et al., 2000; Sirotkin et al., 2000b). Pei and colleagues tested this notion using translation-blocking morpholinos to interfere with *foxH1* (Pei et al., 2007). Morpholino depletion of FoxH1 disrupts epiboly and compromises gastrulation by a mechanism involving regulation of keratin expression (Pei et al., 2007). Difficulty in rescuing the mutant phenotype prevented the authors from determining whether the SMAD Interaction Domain (SID) or another functional domain of FoxH1 is responsible for this activity. In the future, maternal mutants with alleles of *foxH1* producing truncated proteins will be required to corroborate the morpholino phenotype, and to definitively distinguish between maternal and zygotic activities of FoxH1 during gastrulation.

The MZ*sur/foxH1* mutant phenotype is also mild compared to MZ*oep* or compound *sqt*; *cyc* mutant phenotypes, and suggests that additional transcription factors contribute to MZNodal signaling in zebrafish (Gritsman et al., 1999; Pogoda et al., 2000; Sirotkin et al. 2000). Conversely, FoxH1 gain of function suppression of some aspects of the *MZoep* phenotype (e.g., development of axial mesoderm, muscle, trunk defects) in the absence of Nodal signaling indicates that FoxH1 may also have nodal-independent activities (Pogoda et al., 2000). *Mixer/bonnie and clyde* (*bon*) encodes a paired like homeodomain transcription factor that recruits activated SMAD complexes via its SMAD Interaction Motif (SIM) (Randall et al., 2002). In zebrafish, *Bon* expression is partially dependent on *MZoep* and *MZsur* (Kunwar et al., 2003). Compound *MZsur*; *bon* mutants and *MZsur* mutants depleted of *bon* using morpholinos support both overlapping and individual contributions of Sur and Bon to mesendoderm specification (Kunwar et al., 2003). In zebrafish, maternal and zygotic activity of *foxh1* and zygotic activity of *mixer/bon* are essential to maintain the high levels of Nodal signaling required for axial mesoderm development in the early embryo. However, the compound mutant phenotype does not account for all aspects of loss of maternal-zygotic Nodal signaling. These findings provide an example of SMAD-associated transcription factors determining cell-type-specific responses to Nodal signaling, and an indication that additional transcriptional regulators of maternal-zygotic Nodal signaling remain to be discovered.

In *Xenopus* maternal *XSox3*, a Sex-related region Y (SRY)-related high mobility group box transcription factor is asymmetrically distributed along the animal–vegetal axis and is more abundant in the animal region (Zhang et al., 2003). XSox3 is a transcriptional repressor of *nodal related 5* (*Xnr5*) in embryos and animal caps expressing VegT (Zhang et al., 2003). In *Xenopus*, VegT activates *Xnr5* transcription at stages before wholesale embryonic genome activation (Zhang et al., 2003; Yang et al., 2002). The maternal function of XSox3 as an *Xnr5* repressor in embryos was revealed through studies in which C-terminal-directed antibodies were injected into *Xenopus* embryos to prevent XSox3 from binding DNA (Zhang et al., 2003). In related studies, the same antibodies were used to block the zebrafish Sox protein, which caused elevated expression of the *nodal-related 1*, *cyclops*, but not *squint/nodal-related 2* mRNA (Zhang et al., 2004). Suppression of the Sox3 antibody interference phenotypes with Cerberus-short, a secreted BMP and Nodal antagonist, provided additional evidence for a conserved role of maternal XSox3 in limiting the levels of specific Nodals in the early embryo (Zhang et al., 2004). In the future, it will be interesting to investigate the phenotype of Sox3 antibody interference in *MZoep* mutants to determine whether all aspects or only a subset of the interference phenotypes are mediated by elevated Nodal signaling.

VegT is a vegetal pole localized RNA in late-stage oocytes (Heasman et al., 2001). Depletion of Protein or transcripts encoding this T box transcription factor disrupts mesoderm formation (Figure 19) (Heasman et al., 2001). VegT is known to regulate the transcription of several members of the TGF family, including Nodals to promote mesoderm and endoderm development (Heasman et al., 2001; Xanthos et al., 2001). *VegT* transcripts but not its protein are essential for the localization of vegetal localized mRNAs in oocytes (e.g., Vg1 and Wnt11 transcripts), implicating VegT mRNA in maintaining or anchoring these vegetal pole transcripts (Figure 19) (Heasman et al., 2001). The same subset of mRNAs are mislocalized when maternal *Xlsirts* transcripts are depleted, or when Cytokeratin or the endoplasmic reticulum are disrupted; thus, these molecules may represent components of the vegetal anchoring machinery for some vegetal mRNAs in *Xenopus* (Kloc and Etkin, 1994; Alarcon and Elinson, 2001; Heasman et al., 2001). Despite defective localization of *Vg1* and *wnt11* mRNAs in VegT depleted oocytes, supplying *VegT* transcripts later is sufficient to rescue embryonic patterning (Heasman et al., 2001). This suggests that pre-existing oocyte pattern established prior to VegT depletion or any remaining pattern following depletion is sufficient for normal axis formation when VegT is supplied to the embryo before stages when mesoderm induction should occur (Heasman et al., 2001). The dependence of *wnt11* and *Vg1* mRNA localization on VegT places it upstream of *wnt11* and *Vg1* localization. However, the depletion phenotype and rescue experiments suggest VegT protein acts downstream of Wnt11, Vg1, and bicaudal-C (Heasman et al., 2001). Alternatively, localization of *wnt11* and *Vg1* in oocytes is dispensable due to compensation by redundant T box genes. Combinatorial depletion of maternally expressed T box genes is necessary to test this possibility.

Although mutants disrupting the VegT related, *spadetail*, and *brachyury/notail*, have essential zygotic functions in mesoderm patterning, these T box genes are not maternally expressed in the zebrafish (Schulte-Merker et al., 1994; Griffin et al., 1998; Ruvinsky et al., 1998). In contrast, *eomesodermin* is a conserved T box gene whose products localize along the animal–vegetal axis of zebrafish oocytes (Bruce et al., 2003). Eomesodermin has mesoderm inducing activity via a mechanism that is partially dependent on Nodal signaling (Bruce et al., 2003). Abundant maternal protein has hindered attempts to fully eliminate Eomesodermin by morpholino interference (Bruce et al., 2003). A maternal or maternal zygotic mutant disrupting *eomesodermin* will be required to investigate whether it is necessary for establishing the dorsal organizer and inducing mesoderm in the zebrafish.

ACTIVATORS OF VENTRAL FATE: A ROLE FOR PLURIPOTENCY FACTORS IN PROMOTING ZYGOTIC EXPRESSION OF TRANSCRIPTION FACTORS THAT REGULATE VENTRAL FATES

Introducing the nuclear content of somatic cells into oocytes can induce reprogramming and restore pluripotency (Wilmut et al., 1997). This reprogramming activity of oocytes led researchers to pursue the responsible oocyte factors. In 2006, Takahashi and Yamanaka reported the identification of four factors that when expressed together are sufficient to induce pluripotency (Takahashi and Yamanaka, 2006). These pluripotency factors include Oct4 (also known as Pou5f1/Pou2), SRY-related HMG box transcription factor 2 (Sox2), and known tumor suppressors c-Myc, and Krüppel-like factor 4 (Klf4) (Takahashi and Yamanaka, 2006). C-Myc knockout mice are zygotic lethal (Baudino et al., 2002; Davis et al., 1993), and antisense depletion of c-Myc indicates that zygotic c-Myc regulates embryonic cleavages (Paria et al., 1992). Functional redundancy between Myc family members (e.g., N-Myc) makes it challenging to exclude or confirm an essential role for maternal c-Myc in reprogramming (Charron et al., 1992; Malynn et al., 2000). Although *klf4* is detected in oocytes by PCR, functional evidence to support a maternal requirement for Klf4 has not been reported in the mouse (Yan et al., 2010). Zygotic Oct4 and Sox2 are essential for inner cell mass formation during lineage specification in preimplantation stage mouse embryos (Avilion et al., 2003; Nichols et al., 1998; Pesce and Scholer, 2000).

The presence of maternal Oct4 protein in oocytes and embryos through the four-cell stage, and Sox2 protein through implantation stages indicate potential maternal functions for these pluripotency factors in the mouse (Rosner et al., 1990; Avilion et al., 2003; Scholer et al., 1990; Yeom et al., 1996). Although the maternal *sox2* phenotype has not been described, evidence for maternal *sox2* expression comes from maternal-effect Cre activity in the progeny of *sox2Cre* transgenic females (i.e., Cre is active in all progeny independent of inheritance of the *sox2Cre* transgene) (Hayashi et al., 2003). Conditional knockdown of Pou5f1/Pou2/Oct4 in germ cells revealed its essential role in promoting primordial follicle survival; however, premature apoptosis of oocytes prevents analysis of potential later maternal functions (Kehler et al., 2004; Parfenov et al., 2003). Arrest of morpholino-mediated Oct4-depleted mouse embryos prior to blastocyst formation provides support for a contribution of Oct4 to development prior to blastocyst stages (Foygel et al., 2008). Normal development of mismatch control morpholino-treated embryos to blastocyst stage and partial suppression of the Oct4 morpholino arrest phenotypes by *oct4* mRNA indicate specificity. Gene expression profiling of morphants implicates Oct4 in transcriptional and post-transcriptional regulation of gene expression and degradation of maternal mRNAs at the maternal to zygotic transition (Foygel et al., 2008). Cumulatively, these studies indicate maternal contributions for Oct4 and Sox2 pluripotency transcription factors. However, conditional knockouts of Pou5f1/Pou2/Oct4 and Sox2 using a late oocyte promoter will clarify their maternal contributions to preimplantation development in the mouse, and provide an opportunity to directly test whether they are the essential factors mediating reprogramming of somatic cell nuclei in oocytes.

Mutations disrupting zebrafish *spiel-ohne-grenzen* (*spg*)/*pou2* were originally identified based on the midbrain hindbrain patterning phenotypes of zygotic mutants (Schier et al., 1996). Abundant maternal *pou2* transcripts hinted at potential maternal compensation in zygotic *pou2* zebrafish mutants (Burgess, Reim et al. 2002). Initial studies of maternal *pou2* function using morpholinos to target maternal *spg*/*pou2* in the zygotic *spg*/*pou2* mutant background caused developmental arrest prior to gastrulation stages (Burgess et al., 2002). Later studies of zebrafish maternal-zygotic *spg*/*pou2* mutants corroborated a maternal requirement for *spg*/*pou2* during early embryonic development in endoderm formation (Lunde et al., 2004; Reim et al., 2004), regulation of epiboly during gastrulation (Reim and Brand, 2006), and revealed its function as an activator of Bmp2 ligand expression, which is necessary to promote ventral fates (Figure 19) (Reim and Brand 2006). Though *spg/pou2* mRNA localizes asymmetrically along the animal–vegetal oocyte axis, maternal *spg/pou2* is dispensable for oocyte development, formation of the animal–vegetal axis, and egg activation. These findings indicate that maternal *pou2* transcripts are localized in oocytes to position Pou2 protein for its essential maternal role in activating the BMP pathway to promote ventral fate specification.

Morpholino depletion of four *B1 class sox* genes point toward redundant functions of these genes in zebrafish to regulate zygotic expression of *bmp4*, *bmp7*, and Bmp target genes, and, like MZ *spg/pou2*, to promote epiboly (Okuda et al., 2010). In contrast to the mouse, maternal-zygotic mutants and depletion studies do not support a requirement for Pou2 or these B1 class Sox proteins in controlling pluripotency of somatic and germline cells in the zebrafish. However, both in zebrafish and in the mouse *pou2* regulates specification and differentiation of endodermal fates (Lunde et al., 2004; Reim et al., 2004). Due to the early lineage specification defects of mouse *pou2* and *sox2* mutants, it remains unclear whether Pou2/Oct4 and Sox2 have conserved functions in dorsal–ventral patterning, which is regulated by maternal and zygotic contributions in zebrafish and is controlled zygotically in the mouse.

LIGANDS AND TRANSDUCERS OF VENTRAL FATE

Ectopic expression of the zebrafish ortholog of *growth differentiation factor 6*, *gdf6*, a TGFβ family member also known as *radar*, is sufficient to cause ventralization via activation of zygotic *bmp2b/4* and the maternal BMP receptor Alk8 (Goutel et al., 2000; Sidi et al., 2003). Ventralizing activity of Radar in zebrafish mutants requires Bmp2 and Bmp7 ligands, the transcription factor Smad5, and a functional Alk receptor (Goutel et al., 2000; Sidi et al., 2003). Dominant negative forms of Radar and morpholino depletion cause dorsalized phenotypes and demonstrate a necessary role for Radar in promoting zygotic *bmp* expression and BMP signaling activity (Sidi et al., 2003). The dominant negative versions of Radar elicited more severe phenotypes compared to the morpholino induced phenotypes, which could reflect interference of the dominant negative protein with other TGF molecules or an incomplete knockdown of maternal Radar in the morphants. A maternal mutant will be required to distinguish between these possibilities. Similar to Pou2, Radar activates the expression of zygotic genes that are essential for ventral fate specification. These results provide evidence that the cells of the zebrafish embryo are not simply poised to develop as ventral in the absence of signals that promote dorsal fates. Instead, like dorsal specification, the ventral program relies on dedicated maternal inducers of the zygotic components that are required to specify and pattern the embryonic axes.

Zygotic roles of BMP ligands in patterning the embryo have been appreciated in diverse model systems including vertebrates. Dominant negative mutant alleles of *smad5*, also call *somitabun* in zebrafish, cause mild dorsalization of the body axis in zebrafish but are not embryonic lethal; thus, the mutants survive to adulthood and are fertile (Hild et al., 1999; Kramer et al., 2002).

However, the progeny of mutant females are profoundly dorsalized. This maternal-effect *SMAD5* phenotype, and the stronger dorsalized phenotypes of maternal-zygotic *Alk8* receptor mutants implicate a maternal BMP ligand in patterning the zebrafish dorsal–ventral axis (Kramer et al., 2002; Bauer et al., 2001). However, maternal mutants disrupting *Bmp2b* and *Bmp7* indicate that these ligands do not provide essential maternal contributions to patterning (Kishimoto et al., 1997; Dick et al., 2000; Schmid et al., 2000).

Zygotic regulation of dorsal–ventral patterning in zebrafish involves mutual inhibition of dorsal organizer and ventral–posterior activities (reviewed by Schier and Talbot, 2005). Until Radar and Pou2 were identified as maternal inducers of zygotic BMP and ventral patterning genes, ventral was thought to be a default state. Depletion of the *runxbt2* transcription factor causes dorsalized phenotypes, similar to *pou2* and *radar* phenotypes (Flores et al., 2008). However, unlike *radar* and *pou2* phenotypes, dorsalization is independent of Bmp2b, Bmp7, and SMAD5. Instead Runx2bt is necessary to activate the expression of early zygotic transcriptional repressors of dorsal gene expression, namely, *vox*, *vent*, and *ved* (Figure 20) (Flores et al., 2008). Consistent with a role for Runx2bt in promoting ventral fates, excess zebrafish or murine Runx2 causes ventralization (Flores et al., 2008). Although rescue of zebrafish Runx depletion phenotypes by mouse Runx2 suggests that the mouse protein can functionally compensate for depletion of the zebrafish protein knockout of zygotic Runx2 does not cause axis defects in mice (Flores et al., 2008; Hernandez-Gonzalez et al., 2006). *runx2* is expressed in mature oocytes and elsewhere in the ovaries of mice and rats (Park et al., 2010). However, an ovary-specific knockout of Runx2 has not been reported; so, whether Runx2 or a related Runx family member has an essential maternal function in mouse remains to be discovered.

Significant advances have been made in our understanding of the interplay between maternal and zygotic regulators of embryonic axis formation in *Xenopus* and zebrafish (Figure 20). However, the molecular genetic hierarchy required to establish the animal–vegetal axis and conserved asymmetries in early vertebrate oocytes, and the molecular mechanisms linking the oocyte axis to the later developing dorsal–ventral axis remain elusive. Comprehensive analysis of maternal gene function using modern genetic and interference technologies promises to facilitate discovery of unknown regulators of early axis formation in vertebrates.

• • • •

Maternal Control After Zygotic Genome Activation

In the previous chapters we discussed the maternal genes known to contribute to patterning the embryonic axes of fish and frogs, a process that requires maternal inputs and activation of zygotic gene expression. After the zygotic genome is activated, some maternal products avoid degradation. These maternal products continue to function into stages after zygotic genome activation and often cooperate with the *de novo* zygotic products to facilitate normal development. This persisting maternal function can ameliorate zygotic mutant phenotypes when the residual maternal function is sufficient to permit normal development. In some cases, the persisting maternal function compensates entirely, and in others, it delays embryonic arrest until later developmental stages when the maternal product is eventually depleted. The following chapters focus on genes with strict maternal functions at later developmental stages than might be expected, or whose interference or mutant phenotypes are known to be or thought to be attenuated by compensating maternal function.

INSIDE VERSUS OUTSIDE: PATTERNING THE MOUSE BLASTOCYST

The question of whether any pre-pattern present in the oocytes of the mouse influences patterning of the embryo remains an area of active study and controversy. There seems to be agreement and clear evidence for apical basal polarity at the eight-cell stage, after activation of the zygotic genome (for a complete review of the cellular examination of blastocyst formation the reader is referred to (Johnson, 2009). This apical–basal polarity can be seen in the compaction of the eight-cell stage embryo and through the localization of known polarity markers from other organisms and developmental contexts, including partitioning defective 3 (Par-3), partitioning defective 6 (Par-6), and protein kinase C (PKC), but is not detectable during earlier cleavage stages (Pauken and Capco, 1999; Vinot et al., 2005) and reviewed by Muller (2001).

Localized molecules such as integrins, leptin, STAT, and myosin are asymmetrically partitioned during early mitotic divisions (Antczak and Van Blerkom 1997), but so far a polarized maternal gene product essential for generating a well-patterned mouse has not been identified. This has led to models whereby animal–vegetal oocyte or egg asymmetries might serve to bias the system rather than establish a restricted pattern of fates (Gardner, 2007) and reviewed by Zernicka-Goetz (2006). Alternatively, oocyte and egg asymmetries do not have an influence (Alarcon and Marikawa, 2003). Methodologies to deplete maternally supplied gene products promise to provide much needed molecular insight into the maternal and zygotic mechanisms generating lineage restriction and pattern within the mouse blastocyst.

The, subcortical maternal complex, SCMC, is a complex composed of maternal proteins that are essential for developmental progression of the early mouse embryo. The SCMC proteins localize uniformly around the cortex of ovulated mouse eggs and subsequently are excluded from the cell contact surfaces of preimplantation stage mouse embryos by a calcium-mediated mechanism (Tong and Nelson, 1999; Li et al., 2008; Ohsugi et al., 2008; Herr et al., 2008). Eventual localization of the SCMC complex to the outer cells of the compacted morula (i.e., the cells that will develop as trophectoderm; Figure 23) have raised the question of whether the SCMC might contribute to early lineage decisions in the mouse embryo. The phenotypes caused by disrupting individual components of the SCMC implicate the complex in regulating multiple aspects of early cleavage stage development, including the mitotic spindle, specialized cytoskeletal elements, protein translation, and embryonic genome activation (Figure 23). The specific roles identified for SCMC components are discussed in this section.

Mater, Latin for mother, and an acronym for Maternal antigen that embryos require, also known as nucleotide-binding oligomerization domain, leucine-rich repeat and pyrin domain containing 5, Nlrp5, a leucine zipper protein was first identified as a mouse model of premature ovarian failure (Tong and Nelson, 1999). The presence of leucine zipper domains hinted that Mater/Nlrp5 might regulate early embryonic development as a component of a larger protein complex. Diminished abundance of a protein that lacks identifiable domains, Filia, Latin for daughter, but not *filia* transcripts in *mater* mutants indicated that Mater promotes Filia protein abundance or stability (Ohsugi et al., 2008). Maternal-effect *filia* mutants progress further in development compared to *mater* mutants, which arrest at the two-cell stage (Tong et al., 2000a; Tong et al., 2000b; Zheng and Dean, 2009). Filia activity is necessary for spindle morphogenesis as evident by impaired RhoA-dependent recruitment of the mitotic kinase aurora kinase A and compromised RhoA-independent recruitment of polo-like kinase, Plk1, to the spindle poles in *filia* maternal-effect mutants (Zheng and Dean, 2009). Filia protein also regulates localization of the spindle checkpoint protein MAD2l1 to the kinetichore (Zheng and Dean, 2009). The consequences of impaired spindle regulation and checkpoint activation in *filia* mutants are improper segregation of the chromosomes, which results in aneuploidy and developmental arrest. The dependence of Filia abundance on Mater indicates

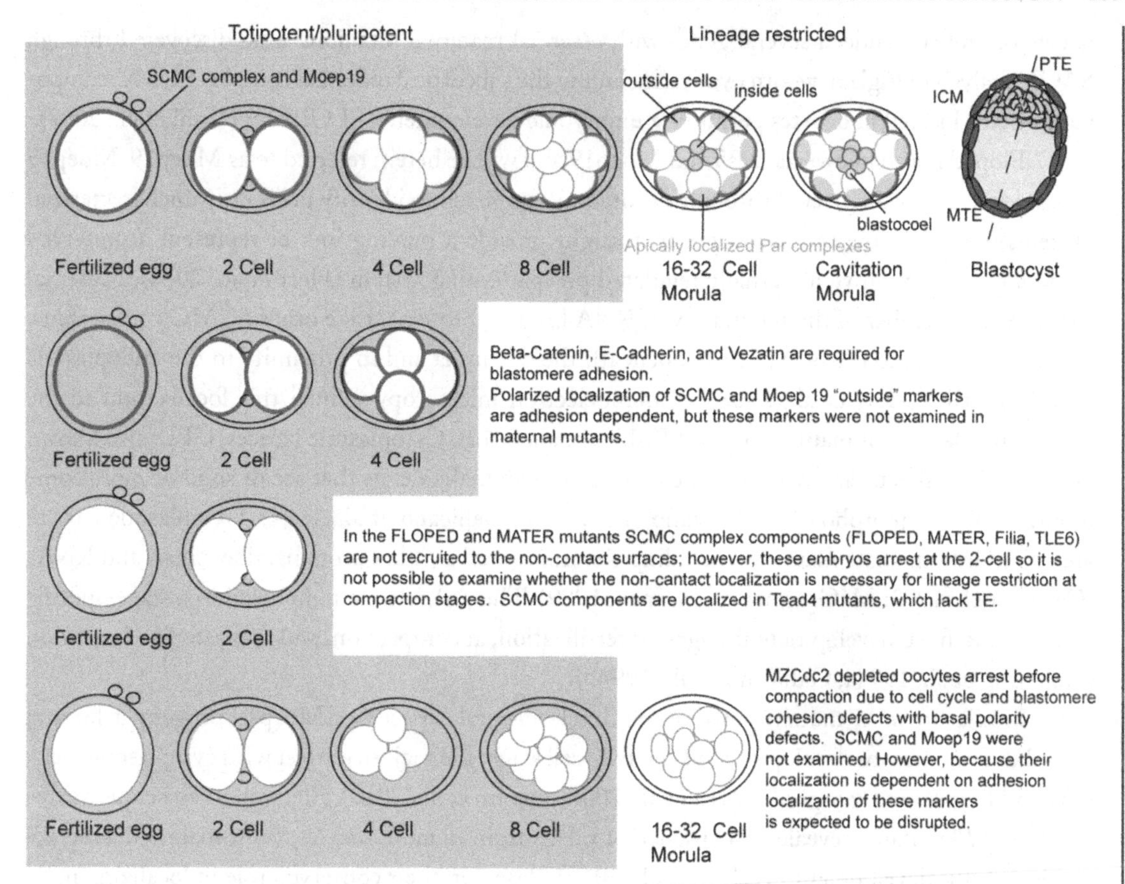

FIGURE 23: Maternal markers and regulators of the first cell lineages in the mouse (i.e., outside: inside or trophectoderm: inner cell mass). Grey marks the stages when the cells are considered to be pluripotent, while the white area marks the transition to lineage commitment. The last stage shown is the stage when the embryos arrest in maternal-effect mutants or depleted embryos. ICM denotes inner cell mass; PTE denotes polar trophectoderm, MTE indicates mural trophectoderm, TE indicates trophectoderm; SCMC indicates the sub-cortical maternal complex.

that developmental arrest of *mater* mutants is likely due in part to spindle defects caused by diminished Filia protein and likely activity.

Factor in the germ line alpha, figla, Italian for daughter, is an oocyte expressed, basic helix–loop–helix (bHLH) transcription factor known to regulate the expression of zona pellucida genes (Liang et al., 1997). Figla is essential for primordial follicle formation (Soyal et al., 2000). As a maternal transcription factor Figla was expected to control the expression of other maternal factors required in the oocyte or embryo. Indeed several genes with essential maternal functions were discovered through SAGE analysis of Figla newborn ovaries, including the subcortical maternal complex SCMC component, Factor located in oocytes permitting embryonic development (FLOPED) (Joshi et al., 2007).

Floped is also known as Ces5 and Moep19 and will be herein referred to as Moep19. Moep19 was independently discovered through a proteomic approach to identify previously uncharacterized proteins present in the mouse oocyte that might provide a missing link or represent a preserved component of a putative maternally regulated pre-patterning system (Herr et al., 2008). Maternal Moep19 is a member of the KH family of RNA binding proteins. Like other SCMC components, Moep19 localizes to the oocyte cortex beneath the oolemma and in proximity to the endoplasmic reticulum (ER) (Herr et al., 2008). Immunoelectron microscopy refined this localization to the cytoplasmic lattices of mature oocytes (Tashiro et al., 2010). Cytoplasmic lattices, CPL, also known as cytoskeletal sheets, are cytoskeletal elements of mammalian eggs that are of solid or more commonly of fibrous morphology (Gallicano et al., 1994; Gallicano et al., 1994). Cytoplasmic lattices are rich in intermediate filaments including Keratin as well as other proteins, ribosomes, and RNAs (Capco et al., 1993; McGaughey and Capco, 1989). These sheets are remodeled in a stage-specific manner at defined developmental stages: at fertilization, at compaction, and at blastodisc formation (Gallicano et al., 1994a; Gallicano et al., 1994b).

Similar to *mater* mutant females, the developmental capacity of Moep19 progeny is limited such that few embryos develop beyond the two-cell stage; the majority arrest with cytoplasmic blebbing and fragmented morphology (Li et al., 2008; Tashiro et al., 2010). Electron microscopic analysis of *moep19* mutants revealed an absence of CPLs from mutant oocytes; thus, maternal Moep19 regulates CPL development (Tashiro et al., 2010). Based on their conserved role in localizing molecules in the oocytes of *Drosophila*, fish, and frogs via interactions with cis-activation sequences, RNA binding proteins would be expected to be key elements of a patterning system. Notably, Moep19 is in the same KH-domain family as MEX-3, which regulates blastomere identity in *C. elegans* (Herr et al., 2008; Draper et al., 1996). The mechanism by which Moep19 mediates CPL formation is not known, but as an RNA binding protein, it is possible that it interacts with or localizes mRNAs to the oocyte cortex analogous to mechanisms localizing mRNAs to the vegetal cortex in *Xenopus* and zebrafish oocytes.

Like Moep19, Peptidylarginine deiminase (Padi6), another component of the SCMC, localizes to cytoplasmic lattices (Esposito et al., 2007). PADI6 regulates CPL formation via a mechanism involving targeting of ribosomes but not, Dnmt1, the methyltransferase discussed earlier, actin, or microfilaments to the cortex (Esposito et al., 2007; Yurttas et al., 2008). Defective CPL formation and ribosomal targeting are associated with a global decline in protein synthesis at the two-cell stage, with only a subset of proteins showing elevated abundance (Esposito et al., 2007). In particular, Spindlin, thought to be a maternal regulator of the meiotic to mitotic transition is more abundant in *Padi6* mutants (Oh et al., 1997; Yurttas et al., 2008). Evidence that Padi6, or an intact CPL, contributes to embryonic genome activation includes impaired protein synthesis associated with reduced RNA Pol II abundance and nuclear translocation, a modest decline in Histone H4K5 acetylation, and reduced activation of the transcription requiring complex in Padi6 maternal effect mutants (Wu et al., 2003). In the future it will be important to determine whether the early defects in lattice formation and ribosome recruitment cause the decline in protein synthesis and later impairment of the EGA, and are also observed in Moep19 mutants. Alternatively, the latter defects may represent independent functions of PADI6.

As with Moep19 and PADI6, immunoelectron microscopy localized Mater to the CPLs (Tashiro et al., 2010). It is not known whether the other SCMC components also localize to the cytoplasmic lattices, but this seems likely as Moep19 and Mater proteins are mutually dependent on one another for their localization, are required to recruit TLE6 to the SCMC, and Mater binds directly to and localizes Filia to the cell cortex (Li et al., 2008) (Figure 23). Restriction of the SCMC to the cells that will eventually develop as trophectoderm raises the possibility that the SCMC contributes to their differentiation as TE. Although the mutants disrupting the SCMC components arrest at the two-cell stage, their maternal proteins perdure and are present in blastocyst stage embryos; therefore it is possible that these proteins have additional functions at these later stages. However, the progeny of most SCMC mutant females arrest at the two-cell stage, before TE and ICM differentiation, thus it is not possible with the existing alleles to investigate whether maternal SCMC, or CPL, localized proteins are necessary for TE specification. The transcription factor Tead4 is zygotically required for trophectoderm formation (Nishioka et al., 2008; Yagi et al., 2007). Evidence that localization of the SCMC is not regulated by Tead4, and SCMC localization is not sufficient to specify trophectoderm fate comes from normal localization of the SCMC to the outer cells of *tead4* mutants (Li et al., 2008). The relationship between SCMC localization, cytoplasmic lattice formation, and trophectoderm specification remains to be resolved.

CDX2: A MATERNAL CONTRIBUTION TO LINEAGE SPECIFICATION IN THE MOUSE

The Cdx2 homeobox transcription factor is required zygotically for trophectoderm formation (Ralston and Rossant, 2008; Strumpf et al., 2005). Earlier phenotypes observed in gain of function and depletion experiments modulating Cdx2 compared to the mouse zygotic knockout hinted at a maternal Cdx2 contribution (Jedrusik et al., 2008). Microarray and RTPCR analysis provided evidence for maternally supplied *cdx2* (Jedrusik et al., 2010). Depletion of maternal Cdx2 using multiple independent approaches revealed an earlier function of Cdx2 in morula stages for apical–basal cell polarity and cell allocation to the TE or ICM lineages (Figure 23) (Jedrusik et al., 2010). Evidence included diminished or absent mRNAs encoding several polarity proteins such as Par3, Par1, apical PKC and E-cadherin in maternal zygotic *cdx2* depleted embryos, while TE markers such as *gata3*, *eomesodermin*, and Troma 1 protein were not detectable (Jedrusik et al., 2010). Elevated levels of *tead4* mRNA hints at potential for feedback regulation, as zygotic Tead4 is required for zygotic expression of *cdx2* (Nishioka et al., 2008; Yagi et al., 2007; Nishioka et al., 2009). How maternal *Cdx2* expression is regulated is not known, but its zygotic regulator, *gata3* has only been detected at the four-cell stage suggesting maternal Cdx2 would be regulated by a Gata3-independent mechanism. Perhaps another Gata transcription factor regulates maternal *cdx2* expression. Another striking difference between the maternal depletion and zygotic mutant phenotypes is the opposite impacts on *nanog* expression; *nanog* is elevated in zygotic mutants but not in the maternal knockdown (Jedrusik et al., 2010; Strumpf et al., 2005). It is possible the lack of *nanog* induction in the maternal knockdown can be attributed to differences in the extent of TE differentiation between maternal and zygotic loss of function. Maternal knockdown of *cdx2* arrests development at an earlier stage, while in the zygotic mutants TE forms but later collapses (Ralston and Rossant, 2008; Strumpf et al., 2005; Jedrusik et al., 2010). Diminished polarity proteins and failure to activate TE marker expression indicate that maternal Cdx2 may regulate TE specification. However, further investigation including analysis of a germline null is needed to understand the degree to which the mechanisms of maternal Cdx2 function in blastocyst formation resemble or diverge from its essential zygotic function as a repressor of ICM markers (fate) during trophectoderm maintenance.

ADHESION IN EPITHELIA, GASTRULATION, AND TISSUE MORPHOGENESIS: G-PROTEIN-COUPLED RECEPTORS

Evidence that maternal G-protein-coupled cell surface receptors regulate cortical actin assembly in *Xenopus* comes from gain of function and maternal depletion studies (Tao et al., 2005; Tao et al., 2007). Maternal depletion of the G-protein-coupled cell surface receptors, XFlop, lysophosphatidic acid 1 (LPA1), or lysophosphatidic acid 2 (LPA2), cause blastula to loose their characteristic morphology and turgidity due to impaired cortical actin assembly independent of microtubules or cytokeratin (Lloyd et al., 2005; Tao et al., 2007; Tao et al., 2005). Impaired blastula integrity and gastrulation are associated with reduced expression of a subset of mesodermal genes without globally altering patterning of the axis (Tao et al., 2005). Reduced levels of C-Cadherin at the cell surface of depleted embryos and suppression of the depletion phenotypes by C-cadherin indicate that these G-protein-coupled receptors mediate cortical actin assembly through regulation of C-cadherin (Tao et al., 2007). Mutant forms of C-cadherin lacking the juxtamembrane region or the p120 Catenin binding motifs support a mechanism whereby C-cadherin mediated actin assembly requires the juxtamembrane region of C-cadherin and its association with p120 Catenin, which is also required for cortical actin assembly (Tao et al., 2007). The ligands activating these intracellular signaling cascades and G-protein coupled receptors to promote cell cohesion and maintain the integrity of the early blastula in *Xenopus* remain to be discovered.

CADHERIN: ROLES IN COHESION AND MORPHOGENESIS

Whether oocyte asymmetries provide instructive information to later establish asymmetries of the developing mouse blastocyst and specify trophectoderm versus inner cell mass lineages remains controversial (reviewed by Johnson, 2009). Interference with E-Cadherin using antibodies impairs compaction and inner cell mass (ICM) formation (Johnson et al., 1986). E-Cadherin homozygous mutants revealed a zygotic requirement for E-Cadherin function during trophectodermal epithelium formation (Larue et al., 1994). The ability of Ecadherin homozygous mutants to undergo compaction and survive to implantation stages was surprising in light of the antibody interference phenotype. However, because the antibody would be expected to interfere with both maternal and zygotic E-cadherin, while the zygotic mutant would only eliminate zygotic cadherin the milder phenotype of the genetic mutant was attributed to compensation by persisting maternal E-cadherin in the blastocyst (Larue et al., 1994). Depleting maternal E-cadherin would thus be expected to result in earlier lethality. Indeed, morpholino depletion of maternal E-cadherin results in embryonic arrest at the two-cell stage, supporting a requirement for maternal E-cadherin in sustaining development to the blastocyst stages in the mutant mice (Kanzler et al., 2003). Injection of mRNA lacking morpholino binding sites was sufficient to rescue the E-cadherin morpholino arrest phenotype indicates specificity (Kanzler et al., 2003). In morpholino depleted mouse embryos β-catenin was not detected at the cell contacts, a phenotype also observed in zygotic E-cadherin null mutants at later stages. Arrest of Ecadherin morphants was not due to altered transcriptional activity based on BrUTP assays (Kanzler et al., 2003), indicating that maternal E-cadherin mediated adhesion is important for development beyond the two-cell stage rather than a change in gene expression.

Consistent with a role for a pre-existing maternal adhesion complex supporting early development in the mouse, conditional E-cadherin knockout, maternal-catenin knockout, or morpholino depletion of a cadherin–catenin complex component, Vezatin, all cause similar cell adhesion and cell division arrest phenotypes (De Vries et al., 2004; Hyenne et al., 2005). Maternal Vezatin is thought to mediate cadherin-dependent intercellular adhesion and suppress lethality in conditional zygotic null *vezatin* mutants until implantation stages (Hyenne et al., 2007); however, the Vezatin maternal mutant or depletion phenotype has not been reported.

In *Xenopus*, maternal depletion indicates early roles for cadherins, EP-cadherin, and their essential adhesion partners, alpha-catenin and plakoglobin, in cell adhesion during early cleavage stages (Heasman et al., 1994; Kofron et al., 1997). Alpha-catenin plays a major role in regulating early cell adhesion as evident by the disaggregation of embryos caused by its depletion from oocytes

(Kofron et al., 1997). Plakoglobin phenotypes are comparatively milder (delayed gastrulation without disrupting patterning) and indicate that it is more important at later stages when zygotically expressed adhesion molecules are also present (Heasman et al., 1994; Kofron et al., 1997). Although plakoglobin and β-catenin appear to be functionally equivalent in some cell culture and gain of function contexts, it is clear from their distinct depletion phenotypes, and the inability of plakoglobin and β-catenin to rescue each other's depletion phenotypes that the maternal functions of these proteins are distinct (Kofron et al., 1997). Future analyses of the cellular phenotype and molecular mechanisms involved will provide insight into the manner in which these cell adhesion molecules distinctly regulate cell adhesion in the early embryo.

ESSENTIAL REGULATORS OF EPIBOLY IN ZEBRAFISH

Expansion of the mouse trophectoderm has been morphogenetically compared to blastodermal cell spreading during zebrafish epiboly (Kane and Adams, 2002). The process of epiboly occurs during gastrulation and involves spreading and thinning of three layers specified at the blastula stage (Figure 24). The most superficial layer is an epithelium known as the enveloping or EVL layer. The EVL is a squamous epithelial monolayer that becomes lineage restricted coincident with ZGA during the midblastula transition (MBT) in zebrafish (Ho, 1992; Kane and Kimmel, 1993; Kimmel et al., 1990). The EVL will develop as the external layer of the epidermis or periderm. Beneath the EVL is the deep layer (DEL), the cells that will form the embryo, while the yolk syncitial layer (a transient extra-embryonic tissue) is most vegetally positioned (Figure 24). Each of these cell layers will undergo movement or spreading toward the vegetal pole until the yolk is completely internalized at the end of gastrulation (Figure 24). Several maternal-effect mutants have been identified through genetic screens in the zebrafish. Although the molecular identity of the disrupted genes remains to be determined for many of these mutants, gene expression and time-lapse analyses reveal cell type-specific requirements for the disrupted genes during epiboly (Figure 24).

Adhesion plays a key role in epiboly of the tissue layers both within and between tissue layers in zebrafish. E-cadherin has been proposed to mediate adhesion between the individual cells within the EVL and between DEL cells and the surface of the EVL (Shimizu et al., 2005; Kane et al., 2005). Maternal-zygotic mutants disrupting *e-cadherin/half-baked* (*hab*) or *epCam*, a type I transmembrane glycoprotein, also known as Tacstd, both impair epiboly within the EVL and compromise epiboly in the deep cell layer by a mechanism that is not fully understood, but involves noncell autonomous interactions between the DEL cells and the surface of the EVL during radial-intercalation (one of the cell behaviors underlying epiboly) of the DEL (Kane et al., 2005; Slanchev et al., 2009; Warga and Kimmel, 1990; Shimizu et al., 2005; Amsterdam and Hopkins, 1999; De Vries et al., 2004; Kane et al., 2005). EpCAM is required in the EVL to mediate cell shape changes and behaviors required for spreading of the EVL in the vegetal half of the embryo (Slanchev et al.,

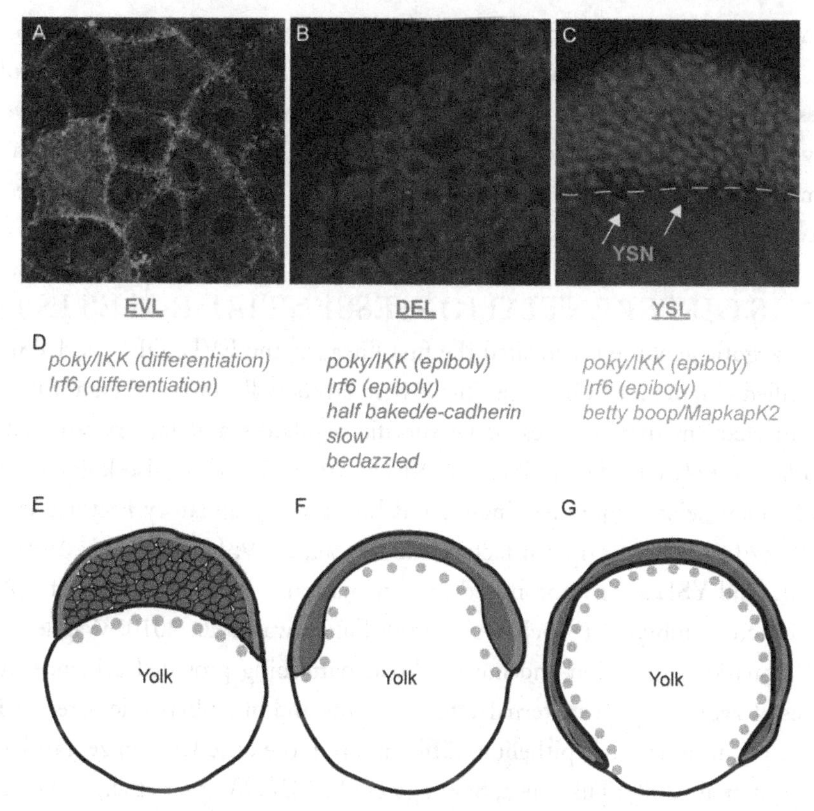

FIGURE 24: Early lineages in zebrafish and regulators of epiboly. (A) The enveloping layer (EVL) marked with membrane tethered GFP. The deep cells labeled with fluorescent dye. (C) The yolk syncitial layer (YSL) nuclei visualized with DAPI. (D) Maternal-effect regulators of epiboly in zebrafish and the tissue where they are required. (E–G) Schematic cross sections depicting movement of the tissues during epiboly. (E) The EVL is the most superficial layer and surrounds the deep cells positioned at the animal pole of the embryo. The Yolk syncitial layer nuclei are positioned at the base of the blastoderm. (F) Mid-epiboly, the EVL, and DEL margins have advanced toward the vegetal pole. The yolk syncitial nuclei move toward the animal pole and at the same time the lateral margins advance toward vegetal as the DEL thins. (G) By late epiboly, the EVL, DEL, and YSL margins are in the vegetal region. Zebrafish maternal-effect mutants with compromised epiboly reveal that movement of these tissue layers is independently regulated, but coordinated as is evident from the genes summarized in panel D.

2009). Expansion of the tight junction domain and diminished E-cadherin in MZEpCam mutants provide evidence that EpCam restricts tight junctions from the basal lateral domain of EVL cells, and promotes E-Cad abundance at the membrane (Slanchev et al., 2009). MZ*epCam* mutants simultaneously depleted of E-Cad lyse during early epiboly. This more severe compound phenotype supports a model whereby EpCam and E-Cad coordinate or cooperate to regulate intercellular adhesion between the EVL and DEL during epiboly.

EPIDERMIS DIFFERENTIATION: ESSENTIAL ROLES IN EPIBOLY

Soon after the zygotic genome is activated the first lineages, the EVL, DEL, and YSL introduced above are specified (Figure 24). Once specified, these layers will move and engulf the yolk during epiboly by a mechanism that involves tissue-specific regulators and factors required for normal epiboly in all layers. *Poky* encodes Inhibitor of NFKB kinase 1, IKK1 (also known as CHUCK; a conserved helix loop helix ubiquitous kinase), and Interferon Regulatory Factor 6 is a maternally and zygotically expressed transcription factor (Sabel et al., 2009; Hatada et al., 1997). Epiboly of the EVL, DEL, and YSL are compromised in *poky* maternal-effect mutants and *irf6* morphants (morpholino injected embryos) (Sabel et al., 2009; Fukazawa et al., 2010; Wagner et al., 2004). Defective EVL marker expression and normal DEL patterning provided evidence that defective morphogenesis of zebrafish *poky* maternal-effect mutants and of embryos depleted of IRF6 is due to a primary defect in superficial epithelium differentiation (i.e., the EVL in zebrafish) (Figure 24) (Sabel, d'Alencon et al., 2009; Fukazawa, Santiago et al., 2010; Wagner et al., 2004). Thus, proper specification and differentiation of the EVL lineage is essential for epiboly of all three germ layers. This is likely due to nonautonomous interactions between the cell types during epiboly. Previous work indicates that microtubule-mediated migration of the YSN, yolk syncitial nuclei, may contribute to epiboly of the EVL by generating forces that contribute to flattening and cell shape changes necessary in the EVL (Betchaku, 1978; Solnica-Krezel, 2006; Solnica-Krezel and Driever, 1994). The EVL cells can nonautonomously influence the DEL by serving as a substrate for the migration of DEL cells during epiboly (Keller and Hardin, 1987; Strahle and Jesuthasan, 1993), or by altering cell adhesion as discussed above for E-cadherin and EpCam and by Zalik et al. (1999). Thus, compromised EVL specification may impact coordinated contraction of microtubules at the margin and thus disrupt epiboly in the underlying cell layers. IKK regulates differentiation of the EVL by a mechanism that requires its kinase activity as mutant phenotypes cannot be rescued by kinase dead versions of IKK; however, the IKK targets essential for EVL differentiation are not known.

In the mouse IKK and IRF6 are essential for differentiation of the embryonic epidermis, skin, craniofacial, and limb development, but whether maternal IRF6 contributes to trophectoderm specification is not known (Ingraham et al., 2006; Richardson et al., 2006; Hu et al., 1999; Takeda et al., 1999; Sil, Maeda et al., 2004). Similarly, dominant negative interference and maternal depletion of IRF6 from oocytes prior to fertilization disrupts differentiation of the superficial epithelium in *Xenopus* (Sabel et al., 2009). Generating IRF6 maternal mutants in zebrafish and mouse and identifying targets of IKK and IRF6 regulation will be important prerequisites to understanding the pathways regulating primary superficial epithelium development in these model systems.

REGULATING EPIBOLY IN THE DEEP CELLS

Mutations disrupting *slow* and *bedazzled* disrupt epiboly of the DEL without impairing patterning or differentiation as is evident from analysis of tissue marker expression (Wagner et al., 2004). Similar to the mutations disrupting *e-cadherin* or *half-baked* discussed earlier, the EVL and YSL complete epiboly while the DEL is delayed in *slow* mutants (Wagner et al., 2004). *bedazzled* mutants show an additional cell shedding phenotype during epiboly, which may indicate a role for *bedazzled* in regulating cell adhesion (Wagner et al., 2004). The molecular identities of *slow* and *bedazzled* when determined will provide missing details of the mechanisms regulating cell-specific and tissue layer-specific movements, and how they are coordinated during morphogenesis.

Unlike Half baked/E-cadherin and the other epiboly regulators discussed above, Map kinase signaling is required to limit actin–myosin contractility to prevent premature blastopore closure as evidenced by the epiboly defects caused by a dominant negative form of p38 and a maternal-effect mutation disrupting its target *betty boop*/*map kap kinase 2* (*MAPKAPK2*) in zebrafish (Figure 25) (Holloway et al., 2009; Wagner et al., 2004). Evidence that *map kap kinase 2* acts within the yolk to prevent premature blastopore closure via phosphorylating an unknown target to limit calcium flux at the margin includes elegant blastoderm yolk recombination experiments, kinase-dependent rescue of the mutant phenotype in a yolk-specific manner, and elevated calcium flux at the margin of mutants preceding premature closure (Figure 25) (Holloway et al., 2009). Previously identified targets of MAPKAPK2 phosphorylation link it to calcium signaling and regulation of actin, including its substrate CREB (leucine zipper transcription factor, cAMP response element binding) in the rat hippocampus (Blanquet et al., 2003; Rane et al., 2001) and substrates Akt and Lsp1p (sphingolipid long-chain base-responsive inhibitors of protein kinases) in neutrophils (Rane et al., 2001; Wu et al., 2007). However, whether these or other as yet discovered targets limit actin contractility during epiboly in zebrafish is not known.

Together these maternal-effect epiboly mutants represent a collection of genes whose products regulate epiboly in a tissue or cell type-specific manner. Future molecular and genetic epistasis analyses will further our understanding of the molecular mechanisms that coordinate differentiation and morphogenesis of these cell layers.

CREB: A ROLE FOR TRANSCRIPTION FACTORS DURING GASTRULATION

The leucine zipper transcription factor, cAMP response element binding (CREB), is highly expressed in *Xenopus* oocytes and is ubiquitous in early zygotes (Sundaram et al., 2003; Lutz et al.,

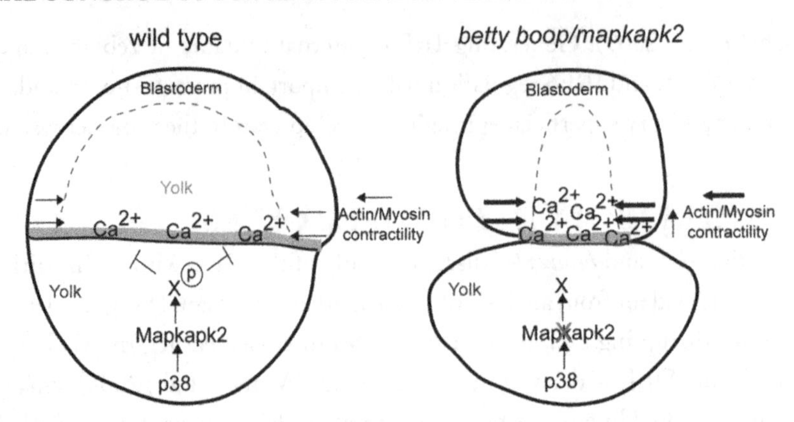

FIGURE 25: Map kinase signaling from the yolk syncitial layer phosphorylates an unknown substrate, which limits calcium abundance and actin myosin contractility at the yolk:blastoderm margin. In *Betty boop* mutants, excess calcium is associated with premature closure of the blastopore ring and rupture of the yolk cell at 50% epiboly. Schematics are lateral views.

1999). However, despite abundant phosphorylated CREB protein in *Xenopus*, maternal CREB is not essential for embryonic development until gastrulation stages when it mediates morphogenesis and ventral mesoderm specification (Sundaram et al., 2003). In morpholino-depleted embryos, zygotic expression of CREB is intact, but it is not sufficient for normal development. Moreover, depletion of zygotic CREB does not exacerbate the maternal CREB depletion phenotype supporting the notion the depletion phenotype is due to impaired maternal function even though the phenotypes manifest at stages after the ZGA (Sundaram et al., 2003). Thus, it seems CREB function in morphogenesis is exclusively regulated by the maternally contributed CREB. The lack of compensation by zygotic CREB suggests that a maternally supplied cofactor may be limiting or no longer present by the time zygotic CREB expression is activated. If a distinct maternal CREB cofactor is present prior to ZGA it should be possible to rescue the maternal CREB depletion phenotype by providing post-ZGA CREB-depleted oocytes with lysate from younger CREB depleted embryos. The putative limiting maternal CREB cofactor awaits identification.

MATERNAL PLANAR CELL POLARITY SIGNALING IN GASTRULATION AND LEFT–RIGHT PATTERNING

In zebrafish, the noncanonical Wnt ligands Silberblick/Wnt11 and Pipetail/Wnt5 discussed earlier do not have essential roles in dorsal–ventral axis patterning, but instead are essential zygotic regulators of individual cell behaviors underlying convergence and extension movements, which narrow and elongate the embryonic body axis and position the germ layers during gastrulation (for a detailed review of the movements of gastrulation, the reader is referred to Yin et al., 2009). The planar cell polarity pathway, as the noncanonical Wnt signaling pathway is also called, is evolutionarily conserved from invertebrates to vertebrates and regulates oriented cell behaviors, reiterating patterns of structures, or their orientation (e.g., hairs or cilia) within the plane of a tissue. Here only the maternal functions of PCP signaling will be discussed (for more information on the contribution of PCP signaling to gastrulation see (Mlodzik, 2002; Roszko et al., 2009; Schier and Talbot, 2005; Tada and Kai, 2009; Wallingford, 2005; Yin et al., 2009).

Where examined thus far most noncanonical or planar cell polarity Wnt signaling components do not have strict maternal functions when zygotic function is intact. Instead the maternal function of these molecules seems to enhance the zygotic function (Ciruna et al., 2002; Topczewski et al., 2001; Wada, Iwasaki et al. 2005).

Loss of maternal and zygotic function of noncanonical Wnt ligands, Silberblick/Wnt11 and Pipetail/Wnt5 produces stronger phenotypes compared to the zygotic loss of function alone, and compound MZ*slb*; *MZppt* mutants show a more severe phenotype, including defective neural tube morphogenesis (Ciruna et al., 2006). Similarly, mutants disrupting maternal and zygotic PCP Wnt pathway components, including the co-receptor *knypek/glypican4*, and a transmembrane PDZ domain-binding protein transducer of PCP Wnt signaling, VanGogh2, also known as *trilobite*, exacerbate the zygotic gastrulation and neurulation phenotypes (Jessen et al., 2002).

Characterization of the MZ*trilobite/vang2* and *MZknypek* phenotypes revealed a previously unappreciated role for Vang2 in patterning the left-right axis, where PCP function asymmetrically positions cilia (Borovina et al., 2010). Other components of the PCP pathway, the cytoskeletal effectors in particular, myosin, Rho, and Rho kinase, have already been discussed for their essential maternal roles in meiosis. However, these proteins are targets of multiple signaling pathways and their function in meiosis is likely PCP independent as meiosis phenotypes have not been reported for other PCP components in zebrafish. In zebrafish, maternal PCP signaling contributes to proper morphogenesis during gastrulation, neural tube formation, and to cilia orientation.

MATERNAL CILIA CONTRIBUTE TO ZYGOTE SIGNALING

Cilia contribute to diverse processes in the developing embryo, including patterning of the axes. The zebrafish *oval* mutant disrupts the intraflagellar transport protein 88, Ift88, and causes ciliary assembly and maintenance defects (Tsujikawa and Malicki, 2004). Cilia form in zygotic *oval* mutants due to maternally supplied *oval* function (Huang and Schier 2009). While maternal *oval* is not strictly required, loss of maternal and zygotic *oval* function ablates cilia, but leaves the basal body intact (Huang and Schier, 2009). The absence of ciliary axonemes specifically impacts zygotic Hedgehog, but not canonical or PCP Wnt signaling (Huang and Schier, 2009).

SCRIBBLE: REGULATING MIGRATION OF GASTRULA CELLS AND BRACHIOMOTOR NEURONS

Landlocked/Scribble is a cytoplasmic Leucine Rich Repeat, PDZ domain protein whose zygotic function is essential for brachiomotor neuron migration (Wada et al., 2005). Zygotic alleles disrupting scribble are viable; thus, making it possible to obtain adults to examine maternal scribble function (Wada et al., 2005). Maternal Scribble 1 but not zygotic Scribble regulates convergence and extension by a mechanism requiring the first PDZ domain of the Scribble protein in zebrafish (Wada et al., 2005). Knockdown of vang2 in MZl and locked/scribble mutants modestly enhanced the convergence extension phenotype, suggesting that scribble and vang2 may interact to regulate gastrulation movements (Wada et al., 2005). Interestingly although Vang2 and Scribble both regulate convergence and extension and brachiomotor neuron migration, Scribble does not seem to be required for neural tube formation. Analysis of the compound maternal and zygotic mutants and investigation of potential genetic interaction or modulation of the MZvang2 phenotype will provide insight into the common and divergent mechanisms underlying cell movements and behaviors in distinct cell types during embryonic development.

LATE OR PERSISTING MATERNAL FUNCTION BEYOND GASTRULATION

In the zebrafish, several mutants show morphological abnormalities only at stages after activation of zygotic gene expression is initiated in wild-type embryos. These mutants display defects in morphogenesis and in cell survival (Wagner et al., 2004); however, until the disrupted genes are identified and the temporal and spatial distribution of their gene products are ascertained it is difficult to distinguish between a stable maternal product that is necessary for advanced developmental stages or a requirement for the maternal product to activate an essential zygotic target necessary after ZGA.

The maternal-effect gene *pollywog* regulates gastrulation movements without impairing patterning (Wagner et al., 2004). The most prominent movement defects are observed in the prechordal plate mesoderm indicating that *pollywog* might function specifically in the prechordal plate or in an adjacent tissue to mediate prechordal plate migration. Molecular identification of the *pollywog* locus and chimeric analysis will distinguish between a cell autonomous and noncell autonomous function in prechordal plate morphogenesis.

Zebrafish *pug* mutants have diminished midbrain–hindbrain patterning, cerebellum, tail, and pectoral fin development (Wagner et al., 2004). *pug* is not required to induce the regional organizer of the midbrain-hindbrain (i.e., the isthmic organizer), or to specify the presomitic mesoderm, but *pug* is required to maintain these patterning centers (Wagner et al., 2004). The molecular identity of the disrupted gene in *pug* mutants has not yet been reported; however, the posterior body morphogenesis and midbrain-hindbrain defects are reminiscent of perturbed Fgf or canonical Wnt signaling phenotypes.

• • • •

Compensation by Stable Maternal Proteins

ROLES FOR MATERNAL REPLICATION COMPLEX AND DNA REPAIR COMPONENTS

Several examples of zygotic mutants in which components of the replication and repair machinery are disrupted have been identified, including *flat head/pol delta1*, *mcm5*, and *topoisomerase 2A* in zebrafish (Ryu et al., 2005; Plaster et al., 2006; Dovey et al., 2009), and *topoisomerase I* and *topoisomerase 2A* in the mouse (Morham et al., 1996; Akimitsu et al., 2003). The ability of the embryo to survive to developmental stages well beyond embryonic genome activation has been attributed to compensation by stable maternally loaded transcripts or protein; however, the maternal mutant phenotypes have not been reported.

MATERNAL CONTRIBUTIONS TO SEGMENTATION AND SOMITE DEVELOPMENT

The somites are reiterated segments that form from mesoderm tissue during development, and will give rise to the axial skeleton and skeletal muscles in vertebrates. In zebrafish somite boundary formation begins with formation of primary epithelial boundaries via accumulation of extracellular matrix in the intersegmental regions (Henry et al., 2005). The *misty somites* mutant was identified through transposon mediated gene trap in zebrafish (Kotani and Kawakami, 2008). The insertion disrupts a conserved vertebrate-specific protein without obvious functional domains (Kotani and Kawakami, 2008). The *misty somites* gene trap allele does not completely eliminate Misty somites activity, and homozygous mutants for the insertion are viable to adulthood (Kotani and Kawakami, 2008). Maternal misty somites function regulates epithelialization during somite boundary formation and later maintains somite boundaries to allow normal muscle development through an unknown pathway and mechanism.

ALDH2: MATERNAL ROLES FOR RETINOIC ACID IN ENDODERM DEVELOPMENT

Retinoic acid is a derivative of vitamin A, retinol. Retinol dehydrogenases convert retinol to retinaldehyde, which is then oxidized by Retinaldehyde dehydrogenases, Raldh or Aldh, to produce Retinoic acid. Aldh2 is the main, but not the only retinaldehyde dehydrogenase expressed in the early embryos of vertebrates. Several mutant alleles disrupting Aldh2, *neckless*, have been identified in zebrafish. These mutants reveal essential roles for zygotic Aldh2 in proper anterior–posterior axis formation, midline mesoderm specification, neural tube and fin development (Gibert et al., 2006; Hamade et al., 2006; Keegan et al., 2005; Begemann et al., 2004; Begemann and Meyer, 2001; Begemann et al., 2001). Treating zygotic mutants with DEAB, a competitive reversible inhibitor of all Aldhs, produced stronger endoderm phenotypes, and suggests that other Aldhs or maternal Aldh2 attenuate the zygotic endoderm phenotype (Alexa et al., 2009). To distinguish between these possibilities Alexa and colleagues used translation and splice blocking morpholinos, which produced phenotypes reminiscent of treatment with DEAB implying that Aldh2 function (presumably maternal) rather than another Aldh was compensating in the endoderm of zygotic mutants.

GART AND PAICS: PURINE SYNTHESIS AND PIGMENTATION

Purines can be produced by a ten-step enzymatic pathway or by recovery or salvage from intracellular turnover. The liver is a major site of *de novo* synthesis, and purines produced in the liver supplement the salvage pathways of cell types that cannot carry out *de novo* purine synthesis. Two mutations disrupting components of the *de novo* purine synthesis pathway are required zygotically for pigment synthesis, retinoblast proliferation, and cell cycle progression during eye growth (Ng et al., 2009). *gart* encodes phosphoribosylglycinamide formyltransferase, phosphoribosylglycinamide synthetase, which catalyzes early steps of inosine monophosphate (IMP) synthesis. *Paics* encodes phosphoribosylaminoimidazole carboxylase, phosphoribosylaminoimidazole succinocarboxamide synthetase, which catalyzes later steps of IMP synthesis. Some homozygous mutants escape the zygotic recessive phenotypes and develop to adulthood possibly due to utilization of the salvage pathway.

The maternal mutant phenotype reveals essential contributions for both enzymes, which cannot be compensated for entirely by zygotic function. The maternal-effect phenotype caused by loss of maternal *paics* is relatively mild compared to the zygotic phenotype; the progeny of mutant mothers have a normal anterior–posterior axis, but reduced pigmentation (Ng et al., 2009). The maternal zygotic phenotype is considerably stronger than the zygotic phenotype caused by *paics* and is characterized by a shorter anterior–posterior axis, reduction of head structures and pigmentation, and cardiac edema (Ng et al., 2009). The maternal-effect and maternal-zygotic phenotypes of *gart* are more severe than the maternal-effect and MZ*paics* phenotypes (Ng et al., 2009). These embryos show developmental delays, anterior defects, lack of pigmentation, and development arrests at 28 hours post-fertilization (Ng et al., 2009). In the future, it will be important to investigate whether the maternal defects are solely due to reduced IMP, or also involve phenotypes caused by the accumulation of substrates of the synthesis pathway. The similar zygotic phenotypes of *gart* and *paics* suggest that these defects are due to compromised IMP availability. The stronger maternal-effect phenotypes of gart indicate that accumulation of substrates may impair embryonic development, as *Gart* mediates earlier steps in the pathway.

• • • •

Maternal Contributions to Germline Establishment or Maintenance

In all animals examined conserved mRNAs, *vasa*, *dazl*, and *nanos* are restricted to a few cells in the embryo, the primordial germ cells, PGCs, which will give rise to the germline. The precise developmental stage when PGCs are specified varies among species (McLaren, 2003; Zhou and King, 2004; Anderson et al., 2007; Braat et al., 1999; De Felici, 2001; Ewen-Campen et al.; Hammond and Matin, 2009; Hashimoto et al., 2004; Slanchev et al., 2005; Weidinger et al., 2003; Wylie, 2000; Anderson et al., 2001; Wylie et al., 1985). In animals that specify PGCs during gastrulation, germ cell specification and survival are zygotically regulated, while animals that specify the germline via inheritance of germ plasm maternal gene products are responsible for specifying the primordial germ cells of the embryo. In these animals, both maternal and zygotic gene products contribute to survival and maintenance of the PGCs. Despite differences in timing and whether maternally or zygotically controlled, the molecules essential for PGC survival and development are conserved. Maternally regulated mRNA localization before fertilization is crucial for germ cell formation in many animals including *Drosophila*, *Xenopus*, and zebrafish. In *Drosophila*, failure to localize germ plasm to the posterior pole of oocytes causes grandchildless phenotypes; the mothers are overtly normal, but their progeny lack PGCs and thus are sterile (Boswell and Mahowald, 1985; Ephrussi and Lehmann, 1992; Lasko, 1992; Lehmann and Nusslein-Volhard, 1986; Rongo et al., 1997). In this chapter, the maternal regulators of PGC specification in vertebrates are described.

Maternal specification of the germline involves producing, positioning, and protecting the germ plasm mRNAs within the oocyte (Figure 26). Post-transcriptional mechanisms regulate mRNA translocation to the correct subcellular location to ensure that the message is protected from degradation and is translated only in the germline of the embryo. The genes and mechanisms that mediate RNA localization in oocytes and embryos are not fully understood; however, they are known to involve RNA and protein complexes (RNPs), cis elements within the 3′UTR and RNA binding proteins recognizing those motifs and acting to promote association with motor proteins, polyadenylation factors, or translational activators or repressors (reviewed by King et al., 2005; Kloc and Etkin, 2005; Minakhina and Steward, 2005).

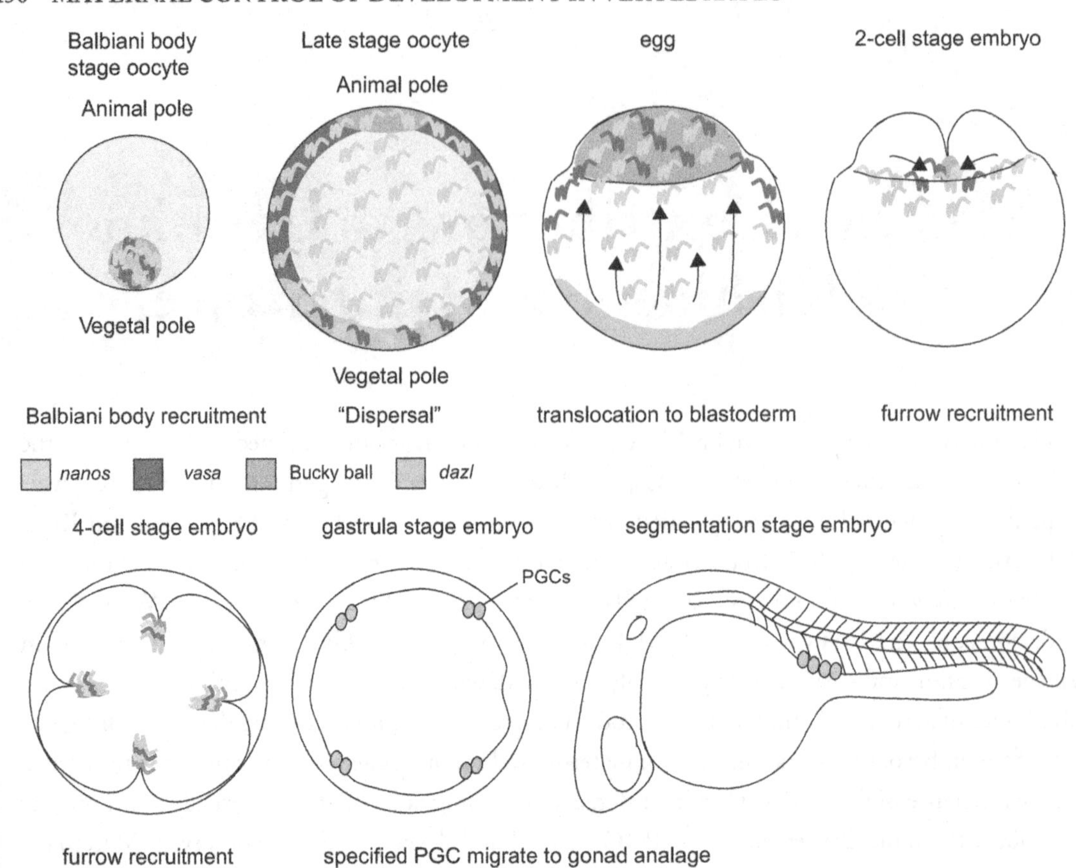

FIGURE 26: In some vertebrates, such as the frogs and zebrafish, germ plasm is set aside in oocytes, first in the Balbiani body, and later remains at the vegetal pole in frogs (not shown) or disperses in late-stage oocytes such that germ plasm transcripts occupy distinct domains along the oocyte axis. When the egg is activated, the germ plasm mRNAs translocates to the blastoderm. During the first two cleavage cycles in fish, the germ plasm is recruited to the furrows. The cells that inherit the germ plasm will develop as the primordial germ cells PGCs. In gastrula stage embryos, the PGCs migrate to the prospective gonad where they will proliferate and eventually give rise to the cells that will develop as eggs or sperm.

FatVg discussed earlier for its maternal contribution to cortical rotation and patterning localizes to vesicles at the vegetal pole of eggs in the vicinity of the germ plasm in *Xenopus*. Its molecular identity and depletion phenotype support a mechanism by which FatVg could regulate vesicular trafficking to ensure proper segregation of the germ plasm as FatVg depletion results in excessive and precocious aggregation of germ plasm associated organelles and mRNAs in *Xenopus* embryos (Chan et al., 2007). However, further studies are needed to functionally demonstrate the extent to which defective trafficking might contribute to germ cell deficiency.

Dazl, Deleted in Azoospermia-like, is a conserved RNA binding protein that functions in germ cell development in many species from insects to man (Houston et al., 1998; Houston and King, 2000; McNeilly et al., 2000; Hashimoto et al., 2004; Ruggiu et al., 1997; Saunders et al., 2003). In *Xenopus* and in zebrafish *dazl* transcripts are localized to the Balbiani body of early-stage oocytes and later localize to the vegetal pole by a mechanism that, in zebrafish, requires Bucky ball mediated Balbiani body formation (Chang et al. 2004; Bontems et al., 2009; Kloc et al., 2001; Kosaka et al., 2007; Maegawa et al., 1999; Marlow and Mullins, 2008; Houston et al., 1998). A conserved role for Dazl in regulating germ cell differentiation and survival is evident from germ cell deficiencies in *Xenopus* embryos depleted of maternal Dazl, in *dazl* mutant mice, and in humans with mutant Dazl genes (Houston and King, 2000; Ruggiu et al., 1997; Saunders et al., 2003). Studies of *dazl* mRNA localization in zebrafish oocytes and embryos identified multiple regions of the 3′UTR that regulate its temporal localization at distinct developmental stages in oocytes and to the cleavage furrow of embryos where it protects a subset of germ plasm mRNAs and promotes their translation (Kosaka et al., 2007; Mishima et al., 2006).

As discussed earlier, miRNAs are key regulators of mRNA turnover at developmental transitions, an activity mediated in part by miRNA sites within the 3′UTR of target mRNAs, including germ plasm mRNAs of zebrafish (Mishima et al., 2006). A key step in germline specification involves selective protection of the germ plasm mRNAs in the prospective germline and their degradation in the soma. Evidence that Dazl and Deadend (Dnd), an RNA binding protein required for PGC survival, specifically protect germ plasm mRNAs from degradation in the germline comes from studies in zebrafish (Mishima et al., 2006; Takeda et al., 2009; Weidinger et al., 2003). Dazl and Dnd are examples of two mechanisms to protect mRNAs from miRNA-mediated degradation; Dazl promotes polyadenylation and translation and Dnd competes with miR430 for binding to target sites within the 3′UTR (Kedde et al., 2007; Mishima et al., 2006).

In wild-type zebrafish embryos, maternally inherited germ plasm components including *vasa*, and *dazl*, and *nanos* accumulate at the cleavage furrows during the first and second cleavage cycles (Figure 26) (Kosaka et al., 2007; Yoon et al., 1997). The accumulation of germ plasm to the cleavage planes only during the first few cleavages suggests that germ plasm recruitment might be coupled to or limited to the early cleavage cycles. However, accumulation of germ plasm in the progeny of *acytokinesis* maternal-effect mutants which show defective cleavage provides evidence that germ plasm recruitment can be uncoupled from cleavage regulation (Kishimoto et al., 2004). Delayed (longer) cleavage cycles and impaired germ plasm accumulation of *cellular atoll/sas6* maternal-effect progeny indicate that there is a timing component to germ plasm recruitment (Yabe et al., 2007). Together, these maternal-effect mutants argue that the temporal window for germ plasm accumulation is restricted to the first cleavage cycles, but is independently regulated from counting the number of cell divisions. In *cellular atoll* mutants, the furrow forms, but at a later time and consequently is either no longer competent or no longer permissive for germ plasm localization. Alternatively, "nonrecruited" germ plasm is no longer available for accumulation. Injecting germ plasm components into these mutants may distinguish between these possibilities. If the cleavage furrows were no longer competent to recruit germ plasm, this would suggest that a maternal protein or complex necessary for furrow competence becomes limiting after the second cleavage cycle.

Whether a particular germ plasm component or a combination of these maternally supplied factors is essential to specify the germline and the order in which germ plasm components are recruited to the cleavage furrows in zebrafish awaits identification of maternal mutants disrupting each germ plasm component. Without maternal Bucky ball *dazl*, *nanos*, and all other germ plasm mRNAs examined are not maintained in oocytes; therefore, these germ plasm mRNAs are not available to the embryo (Marlow and Mullins, 2008; Bontems et al., 2009). In *bucky ball* mutants, defective germ plasm recruitment in embryos is secondary to defective animal–vegetal polarity in primary oocytes, which leads to failure to localize and maintain the germ plasm mRNAs, polyspermy, and defective cleavage patterns in fertilized embryos (Bontems et al., 2009; Marlow and Mullins, 2008). However, Bucky ball fusion proteins do localize to the cleavage furrows of early wild-type embryos (Bontems et al., 2009). Localization of Buc to the furrow would be consistent with a potential later role for Bucky ball in recruiting germ plasm in the embryo, possibly in an analogous manner to its recruitment of germ plasm mRNAs to the Balbiani body of oocytes. A necessary role for *bucky ball* in directing germ plasm the embryo cannot be confirmed or excluded with the existing mutant alleles since they impair Balbiani body assembly in primary oocytes and disrupt animal–vegetal axis development (Bontems et al., 2009; Marlow and Mullins, 2008).

Zebrafish embryos depleted of primordial germ cells, PGCs, develop as sterile males exclusively; thus, PGCs or signals produced by the primordial germ cells are obligatory for development of the female gonad (Ciruna et al., 2002; Draper et al., 2007; Houwing et al., 2007; Slanchev et al., 2005). When maternally provided to the early embryo, nanos permits PGC survival and establishment of the gonad in mutants lacking zygotic *nanos* function; thus, maternal nanos is dispensable for differentiation and development of late-stage oocytes or for maturation or fertilization (Draper et al., 2007). However, later in the adult defective germline stem cell, renewal causes progressive depletion of the germline stem cells and female sterility (Draper et al., 2007). The limited progeny that are produced from *nanos* mutant females are morphologically normal, except they fail to maintain PGCs due to the absence of maternal *nanos* function, and will develop as sterile males (Draper et al., 2007). Thus, in zebrafish, maternal nanos is not required to specify the germline, but rather is required to maintain it by a mechanism that is not fully understood.

• • • •

Perspective

Although significant progress has been made toward deciphering the contribution of maternal products to vertebrate development, there are several indicators that we are only beginning to discover how early development is regulated. Evidence for this notion includes the vast number of gene products that are maternally expressed in vertebrates whose maternal functions have not been experimentally explored genetically or by interference technology. Furthermore, the maternal-effect screens that have been conducted have not reached saturation; for many processes highlighted in this review, only a single or a couple of essential genes have been identified, and the molecular identity of the majority of disrupted genes from forward maternal-effect screens remains unknown. For those processes, where multiple genes have been identified, such as zygotic genome activation and morphogenesis, it is not clear if these genes act in one pathway or represent multiple pathways regulating the same process. Analysis of compound mutants as has been conducted for dorsal–ventral axis formation will be required to distinguish between these possibilities and to order pathway components. It is clear that large gaps in our understanding of the mechanisms and molecular pathways involved remain to be filled even for the best-understood maternally regulated processes.

In the coming years, genetic and targeted depletion are anticipated to continue to illuminate maternal control of developmental processes essential both for embryonic development but also for fertility. It is likely that these approaches alone will not be sufficient because many of the pathways and molecules highlighted in this review, in particular, Bmps, Wnts, the small GTPases, cytoskeletal components, and cell cycle regulators, are known to function at multiple time points during oogenesis, maternally during embryogenesis, and later zygotically in embryos or in the adult animal. The emergence of conditional systems and rescue-mediated bypass of zygotic function has and will continue to facilitate conditional elimination of gene function so a gene's potential maternal function is tangible. Improved access to maternally regulated processes will fill gaps in our current understanding of the mechanisms and pathways responsible for maternally controlled aspects of development and will facilitate discovery of novel contributors to early embryonic development and fertility, including previously unknown pathways and still missing components of known pathways.

Acknowledgments

I would like to thank Andreas Jenny, Sophie Rothhämel, and Alex Schier for helpful discussions and comments on the manuscript. Work on maternal control of oocyte polarity and animal–vegetal axis formation in my laboratory is supported by the National Institutes of Health GM089979.

References

Abdelilah, S., and Driever, W. (1997). Pattern formation in janus-mutant zebrafish embryos. *Dev Biol* 184, pp. 70–84.

Abdelilah, S., Solnica-Krezel, L., Stainier, D.Y., and Driever, W. (1994). Implications for dorsoventral axis determination from the zebrafish mutation janus. *Nature* 370, pp. 468–71.

Abrams, E.W., and Mullins, M.C. (2009). Early zebrafish development: it's in the maternal genes. *Curr Opin Genet Dev* 19, pp. 396–403.

Abrieu, A., Doree, M., and Fisher, D. (2001). The interplay between cyclin-BCdc2 kinase (MPF) and MAP kinase during maturation of oocytes. *J Cell Sci* 114, pp. 257–67.

Aiken, C.E., Swoboda, P.P., Skepper, J.N., and Johnson, M.H. (2004). The direct measurement of embryogenic volume and nucleo-cytoplasmic ratio during mouse pre-implantation development. *Reproduction* 128, pp. 527–35.

Akimitsu, N., Kamura, K., Tone, S., Sakaguchi, A., Kikuchi, A., Hamamoto, H., and Sekimizu, K. (2003). Induction of apoptosis by depletion of DNA topoisomerase IIalpha in mammalian cells. *Biochem Biophys Res Commun* 307, pp. 301–7.

Akkers, R.C., van Heeringen, S.J., Jacobi, U.G., Janssen-Megens, E.M., Francoijs, K.J., Stunnenberg, H.G., and Veenstra, G.J. (2009). A hierarchy of H3K4me3 and H3K27me3 acquisition in spatial gene regulation in *Xenopus* embryos. *Dev Cell* 17, pp. 425–34.

Alarcon, V.B., and Elinson, R.P. (2001). RNA anchoring in the vegetal cortex of the *Xenopus* oocyte. *J Cell Sci* 114, pp. 1731–41.

Alarcon, V.B., and Marikawa, Y. (2003). Deviation of the blastocyst axis from the first cleavage plane does not affect the quality of mouse postimplantation development. *Biol Reprod* 69, pp. 1208–12.

Alexa, K., Choe, S.K., Hirsch, N., Etheridge, L., Laver, E., and Sagerstrom, C.G. (2009). Maternal and zygotic aldh1a2 activity is required for pancreas development in zebrafish. *PLoS One* 4, p. e8261.

Alizadeh, Z., Kageyama, S., and Aoki, F. (2005). Degradation of maternal mRNA in mouse embryos: selective degradation of specific mRNAs after fertilization. *Mol Reprod Dev* 72, pp. 281–90.

Allard, P., Champigny, M.J., Skoggard, S., Erkmann, J.A., Whitfield, M.L., Marzluff, W.F., and Clarke, H.J. (2002). Stem-loop binding protein accumulates during oocyte maturation and is not cell-cycle-regulated in the early mouse embryo. *J Cell Sci* 115, pp. 4577–86.

Anderson, O., Heasman, J., and Wylie, C. (2001). Early events in the mammalian germ line. *Int Rev Cytol* 203, pp. 215–30.

Anderson, R.A., Fulton, N., Cowan, G., Coutts, S., and Saunders, P.T. (2007). Conserved and divergent patterns of expression of DAZL, VASA and OCT4 in the germ cells of the human fetal ovary and testis. *BMC Dev Biol* 7, p. 136.

Antczak, M., and Van Blerkom, J. (1997). Oocyte influences on early development: the regulatory proteins leptin and STAT3 are polarized in mouse and human oocytes and differentially distributed within the cells of the preimplantation stage embryo. *Mol Hum Reprod* 3, pp. 1067–86.

Aoki, F., Worrad, D.M., and Schultz, R.M. (1997). Regulation of transcriptional activity during the first and second cell cycles in the preimplantation mouse embryo. *Dev Biol* 181, pp. 296–307.

Amsterdam, A., and Hopkins, N. (1999). Retrovirus-mediated insertional mutagenesis in zebrafish. *Methods Cell Biol* 60, pp. 87–98.

Aravin, A.A., and Bourc'his, D. (2008). Small RNA guides for de novo DNA methylation in mammalian germ cells. *Genes Dev* 22, pp. 970–75.

Arnold, D.R., Francon, P., Zhang, J., Martin, K., and Clarke, H.J. (2008). Stem-loop binding protein expressed in growing oocytes is required for accumulation of mRNAs encoding histones H3 and H4 and for early embryonic development in the mouse. *Dev Biol* 313, pp. 347–58.

Avilion, A.A., Nicolis, S.K., Pevny, L.H., Perez, L., Vivian, N., and Lovell-Badge, R. (2003). Multipotent cell lineages in early mouse development depend on SOX2 function. *Genes Dev* 17, pp. 126–40.

Bachvarova, R., and De Leon, V. (1980). Polyadenylated RNA of mouse ova and loss of maternal RNA in early development. *Dev Biol* 74, pp. 1–8.

Barkoff, A.F., Dickson, K.S., Gray, N.K., and Wickens, M. (2000). Translational control of cyclin B1 mRNA during meiotic maturation: coordinated repression and cytoplasmic polyadenylation. *Dev Biol* 220, pp. 97–109.

Barrett, S.L., and Albertini, D.F. (2007). Allocation of gamma-tubulin between oocyte cortex and meiotic spindle influences asymmetric cytokinesis in the mouse oocyte. *Biol Reprod* 76, pp. 949–57.

Barrow, J.R. (2006). Wnt/PCP signaling: a veritable polar star in establishing patterns of polarity in embryonic tissues. *Semin Cell Dev Biol* 17, p. 185–93.

Baudino, T.A., McKay, C., Pendeville-Samain, H., Nilsson, J.A., Maclean, K.H., White, E.L., Davis, A.C., Ihle, J.N., and Cleveland, J.L. (2002). c-Myc is essential for vasculogenesis and angiogenesis during development and tumor progression. *Genes Dev* 16, pp. 2530–43.

Bauer, H., Lele, Z., Rauch, G.J., Geisler, R., and Hammerschmidt, M. (2001). The type I serine/threonine kinase receptor Alk8/Lost-a-fin is required for Bmp2b/7 signal transduction during dorsoventral patterning of the zebrafish embryo. *Development* 128, pp. 849–58.

Becker, K.A., and Hart, N.H. (1999). Reorganization of filamentous actin and myosin-II in zebrafish eggs correlates temporally and spatially with cortical granule exocytosis. *J Cell Sci* 112 (Pt 1), pp. 97–110.

Begemann, G., Marx, M., Mebus, K., Meyer, A., and Bastmeyer, M. (2004). Beyond the neckless phenotype: influence of reduced retinoic acid signaling on motor neuron development in the zebrafish hindbrain. *Dev Biol* 271, pp. 119–29.

Begemann, G., and Meyer, A. (2001). Hindbrain patterning revisited: timing and effects of retinoic acid signalling. *Bioessays* 23, pp. 981–86.

Belenkaya, T.Y., Han, C., Standley, H.J., Lin, X., Houston, D.W., and Heasman, J. (2002). pygopus Encodes a nuclear protein essential for wingless/Wnt signaling. *Development* 129, pp. 4089–101.

Bell, A.C., West, A.G., and Felsenfeld, G. (1999). The protein CTCF is required for the enhancer blocking activity of vertebrate insulators. *Cell* 98, pp. 387–96.

Bellipanni, G., Varga, M., Maegawa, S., Imai, Y., Kelly, C., Myers, A.P., Chu, F., Talbot, W.S., and Weinberg, E.S. (2006). Essential and opposing roles of zebrafish {beta}-catenins in the formation of dorsal axial structures and neurectoderm. *Development* 133, pp. 1299–309.

Bennett, J.T., Stickney, H.L., Choi, W.Y., Ciruna, B., Talbot, W.S., and Schier, A.F. (2007). Maternal nodal and zebrafish embryogenesis. *Nature* 450, E1-2; discussion E2–4.

Bernstein, E., Caudy, A.A., Hammond, S.M., and Hannon, G.J. (2001). Role for a bidentate ribonuclease in the initiation step of RNA interference. *Nature* 409, pp. 363–6.

Bernstein, B.E., Mikkelsen, T.S., Xie, X., Kamal, M., Huebert, D.J., Cuff, J., Fry, B., Meissner, A., Wernig, M., Plath, K., et al. (2006). A bivalent chromatin structure marks key developmental genes in embryonic stem cells. *Cell* 125, pp. 315–26.

Besse, F., Lopez de Quinto, S., Marchand, V., Trucco, A., and Ephrussi, A. (2009). *Drosophila* PTB promotes formation of high-order RNP particles and represses oskar translation. *Genes Dev* 23, pp. 195–207.

Betchaku, T.a.T., J.P. (1978). Contact relations, surface activity, and cortical microfilaments of marginal cells of the enveloping layer and of the yolk syncytial and yolk cytoplasmic layers of fundulus before and during epiboly. *J Exp Zool* 206, pp. 381–426.

Betley, J.N., Heinrich, B., Vernos, I., Sardet, C., Prodon, F., and Deshler, J.O. (2004). Kinesin II mediates Vg1 mRNA transport in *Xenopus* oocytes. *Curr Biol* 14, pp. 219–24.

Betschinger, J., and Knoblich, J.A. (2004). Dare to be different: asymmetric cell division in *Drosophila*, *C. elegans* and vertebrates. *Curr Biol* 14, pp. R674–85.

Bierkamp, C., Luxey, M., Metchat, A., Audouard, C., Dumollard, R., and Christians, E. (2010).

Lack of maternal Heat Shock Factor 1 results in multiple cellular and developmental defects, including mitochondrial damage and altered redox homeostasis, and leads to reduced survival of mammalian oocytes and embryos. *Dev Biol* 339, pp. 338–53.

Billett, F.S., and Adam, E. (1976). The structure of the mitochondrial cloud of *Xenopus laevis* oocytes. *J Embryol Exp Morphol* 36, pp. 697–710.

Birsoy, B., Berg, L., Williams, P.H., Smith, J.C., Wylie, C.C., Christian, J.L., and Heasman, J. (2005). XPACE4 is a localized pro-protein convertase required for mesoderm induction and the cleavage of specific TGFbeta proteins in *Xenopus* development. *Development* 132, pp. 591–602.

Birsoy, B., Kofron, M., Schaible, K., Wylie, C., and Heasman, J. (2006). Vg 1 is an essential signaling molecule in *Xenopus* development. *Development* 133, pp. 15–20.

Blackshear, P.J. (2002). Tristetraprolin and other CCCH tandem zinc-finger proteins in the regulation of mRNA turnover. *Biochem Soc Trans* 30, pp. 945–52.

Blanquet, P.R., Mariani, J., and Derer, P. (2003). A calcium/calmodulin kinase pathway connects brain-derived neurotrophic factor to the cyclic AMP-responsive transcription factor in the rat hippocampus. *Neuroscience* 118, pp. 477–90.

Bogard, N., Lan, L., Xu, J., and Cohen, R.S. (2007). Rab11 maintains connections between germline stem cells and niche cells in the *Drosophila* ovary. *Development* 134, pp. 3413–8.

Bolivar, J., Huynh, J.R., Lopez-Schier, H., Gonzalez, C., St Johnston, D., and Gonzalez-Reyes, A. (2001). Centrosome migration into the *Drosophila* oocyte is independent of BicD and egl, and of the organisation of the microtubule cytoskeleton. *Development* 128, pp. 1889–97.

Bolton, V.N., Oades, P.J., and Johnson, M.H. (1984). The relationship between cleavage, DNA replication, and gene expression in the mouse 2-cell embryo. *J Embryol Exp Morphol* 79, pp. 139–63.

Bontems, F., Stein, A., Marlow, F., Lyautey, J., Gupta, T., Mullins, M.C., and Dosch, R. (2009). Bucky ball organizes germ plasm assembly in zebrafish. *Curr Biol* 19, pp. 414–22.

Borovina, A., Superina, S., Voskas, D., and Ciruna, B. (2010). Vangl2 directs the posterior tilting and asymmetric localization of motile primary cilia. *Nat Cell Biol* 12, pp. 407–12.

Bortvin, A., Goodheart, M., Liao, M., and Page, D.C. (2004). Dppa3/Pgc7/stella is a maternal factor and is not required for germ cell specification in mice. *BMC Dev Biol* 4, p. 2.

Boswell, R.E., and Mahowald, A.P. (1985). tudor, a gene required for assembly of the germ plasm in *Drosophila melanogaster*. *Cell* 43, pp. 97–104.

Bouniol, C., Nguyen, E., and Debey, P. (1995). Endogenous transcription occurs at the 1-cell stage in the mouse embryo. *Exp Cell Res* 218, pp. 57–62.

Braat, A.K., Speksnijder, J.E., and Zivkovic, D. (1999). Germ line development in fishes. *Int J Dev Biol* 43, pp. 745–60.

Branford, W.W., and Yost, H.J. (2002). Lefty-dependent inhibition of Nodal- and Wnt-responsive organizer gene expression is essential for normal gastrulation. *Curr Biol* 12, pp. 2136–41.

Brennecke, J., Aravin, A.A., Stark, A., Dus, M., Kellis, M., Sachidanandam, R., and Hannon, G.J. (2007). Discrete small RNA-generating loci as master regulators of transposon activity in *Drosophila*. *Cell* 128, pp. 1089–103.

Bringmann, H. (2008). Mechanical and genetic separation of aster- and midzone positioned cytokinesis. *Biochem Soc Trans* 36, pp. 381–83.

Bringmann, H., Cowan, C.R., Kong, J., and Hyman, A.A. (2007). LET-99, GOA 1/GPA-16, and GPR-1/2 are required for aster-positioned cytokinesis. *Curr Biol* 17, pp. 185–91.

Bringmann, H., and Hyman, A.A. (2005). A cytokinesis furrow is positioned by two consecutive signals. *Nature* 436, pp. 731–34.

Bringmann, H., Skiniotis, G., Spilker, A., Kandels-Lewis, S., Vernos, I., and Surrey, T. (2004). A kinesin-like motor inhibits microtubule dynamic instability. *Science* 303, pp. 1519–22.

Bruce, A.E., Howley, C., Zhou, Y., Vickers, S.L., Silver, L.M., King, M.L., and Ho, R.K. (2003). The maternally expressed zebrafish T-box gene eomesodermin regulates organizer formation. *Development* 130, pp. 5503–17.

Bukovsky, A., Caudle, M.R., Svetlikova, M., and Upadhyaya, N.B. (2004). Origin of germ cells and formation of new primary follicles in adult human ovaries. *Reprod Biol Endocrinol* 2, p. 20.

Bultman, S., Gebuhr, T., Yee, D., La Mantia, C., Nicholson, J., Gilliam, A., Randazzo, F., Metzger, D., Chambon, P., Crabtree, G., et al. (2000). A Brg1 null mutation in the mouse reveals functional differences among mammalian SWI/SNF complexes. *Mol Cell* 6, pp. 1287–95.

Bultman, S.J., Gebuhr, T.C., Pan, H., Svoboda, P., Schultz, R.M., and Magnuson, T. (2006). Maternal BRG1 regulates zygotic genome activation in the mouse. pp. *Genes Dev* 20, pp. 1744–54.

Burgess, S., Reim, G., Chen, W., Hopkins, N., and Brand, M. (2002). The zebrafish spiel-ohne-grenzen (spg) gene encodes the POU domain protein Pou2 related to mammalian Oct4 and is essential for formation of the midbrain and hindbrain, and for pre-gastrula morphogenesis. *Development* 129, pp. 905–16.

Burglin, T.R., Mattaj, I.W., Newmeyer, D.D., Zeller, R., and De Robertis, E.M. (1987). Cloning of nucleoplasmin from *Xenopus laevis* oocytes and analysis of its developmental expression. *Genes Dev* 1, pp. 97–107.

Burns, K.H., Viveiros, M.M., Ren, Y., Wang, P., DeMayo, F.J., Frail, D.E., Eppig, J.J., and Matzuk, M.M. (2003). Roles of NPM2 in chromatin and nucleolar organization in oocytes and embryos. *Science* 300, pp. 633–6.

Cao, R., Wang, L., Wang, H., Xia, L., Erdjument-Bromage, H., Tempst, P., Jones, R.S., and Zhang, Y. (2002). Role of histone H3 lysine 27 methylation in Polycomb-group silencing. *Science* 298, pp. 1039–43.

Capco, D.G., Gallicano, G.I., McGaughey, R.W., Downing, K.H., and Larabell, C.A. (1993). Cytoskeletal sheets of mammalian eggs and embryos: a lattice-like network of intermediate filaments. *Cell Motil Cytoskeleton* 24, pp. 85–99.

Carballo, E., Lai, W.S., and Blackshear, P.J. (1998). Feedback inhibition of macrophage tumor necrosis factor-alpha production by tristetraprolin. pp. *Science* 281, pp. 1001–5.

Carmell, M.A., Girard, A., van de Kant, H.J., Bourc'his, D., Bestor, T.H., de Rooij, D.G., and Hannon, G.J. (2007). MIWI2 is essential for spermatogenesis and repression of transposons in the mouse male germline. *Dev Cell* 12, pp. 503–14.

Carrera, I., Janody, F., Leeds, N., Duveau, F., and Treisman, J.E. (2008). Pygopus activates Wingless target gene transcription through the mediator complex subunits Med12 and Med13. *Proc Natl Acad Sci U S A* 105, pp. 6644–9.

Casanueva, M.O., and Ferguson, E.L. (2004). Germline stem cell number in the *Drosophila* ovary is regulated by redundant mechanisms that control Dpp signaling. *Development* 131, pp. 1881–90.

Castillo, A., and Justice, M.J. (2007). The kinesin related motor protein, Eg5, is essential for maintenance of pre-implantation embryogenesis. *Biochem Biophys Res Commun* 357, pp. 694–99.

Cha, S.W., Tadjuidje, E., Tao, Q., Wylie, C., and Heasman, J. (2008). Wnt5a and Wnt11 interact in a maternal Dkk1-regulated fashion to activate both canonical and non-canonical signaling in *Xenopus* axis formation. *Development* 135, pp. 3719–29.

Cha, S.W., Tadjuidje, E., White, J., Wells, J., Mayhew, C., Wylie, C., and Heasman, J. (2009). Wnt11/5a complex formation caused by tyrosine sulfation increases canonical signaling activity. *Curr Biol* 19, pp. 1573–80.

Chan, A.P., Kloc, M., Larabell, C.A., LeGros, M., and Etkin, L.D. (2007). The maternally localized RNA fatvg is required for cortical rotation and germ cell formation. *Mech Dev* 124, pp. 350–63.

Chang, P., Torres, J., Lewis, R.A., Mowry, K.L., Houliston, E., and King, M.L. (2004). Localization of RNAs to the mitochondrial cloud in *Xenopus* oocytes through entrapment and association with endoplasmic reticulum. *Mol Biol Cell* 15, pp. 4669–81.

Chao, W., Huynh, K.D., Spencer, R.J., Davidow, L.S., and Lee, J.T. (2002). CTCF, a candidate trans-acting factor for X-inactivation choice. *Science* 295, pp. 345–7.

Charron, J., Malynn, B.A., Fisher, P., Stewart, V., Jeannotte, L., Goff, S.P., Robertson, E.J., and Alt, F.W. (1992). Embryonic lethality in mice homozygous for a targeted disruption of the N-myc gene. *Genes Dev* 6, pp. 2248–57.

Cheloufi, S., Dos Santos, C.O., Chong, M.M., and Hannon, G.J. (2010). A dicer-independent miRNA biogenesis pathway that requires Ago catalysis. *Nature* 465, pp. 584–89.

Chen, D., and McKearin, D. (2003). Dpp signaling silences bam transcription directly to establish asymmetric divisions of germline stem cells. *Curr Biol* 13, pp. 1786–91.

Chen, D., and McKearin, D. (2005). Gene circuitry controlling a stem cell niche. *Curr Biol* 15, pp. 179–84.

Chen, H.J., Lin, C.M., Lin, C.S., Perez-Olle, R., Leung, C.L., and Liem, R.K. (2006). The role of microtubule actin cross-linking factor 1 (MACF1) in the Wnt signaling pathway. *Genes Dev* 20, pp. 1933–45.

Cheng, A.M., Thisse, B., Thisse, C., and Wright, C.V. (2000). The lefty-related factor Xatv acts as a feedback inhibitor of Nodal signaling in mesoderm induction and L-R axis development in *Xenopus*. *Development* 127, pp. 1049–61.

Cheng, S.K., Olale, F., Bennett, J.T., Brivanlou, A.H., and Schier, A.F. (2003). EGF-CFC proteins are essential coreceptors for the TGF-beta signals Vg1 and GDF1. *Genes Dev* 17, pp. 31–6.

Chernukhin, I., Shamsuddin, S., Kang, S.Y., Bergstrom, R., Kwon, Y.W., Yu, W., Whitehead, J., Mukhopadhyay, R., Docquier, F., Farrar, D., et al. (2007). CTCF interacts with and recruits the largest subunit of RNA polymerase II to CTCF target sites genome-wide. *Mol Cell Biol* 27, pp. 1631–48.

Chien, A.J., Conrad, W.H., and Moon, R.T. (2009). A Wnt survival guide: from flies to human disease. *J Invest Dermatol* 129, pp. 1614–27.

Christians, E., Davis, A.A., Thomas, S.D., and Benjamin, I.J. (2000). Maternal effect of Hsf1 on reproductive success. *Nature* 407, pp. 693–94.

Cifuentes, D., Xue, H., Taylor, D.W., Patnode, H., Mishima, Y., Cheloufi, S., Ma, E., Mane, S., Hannon, G.J., Lawson, N.D., et al. (2010). A novel miRNA processing pathway independent of Dicer requires Argonaute2 catalytic activity. *Science* 328, pp. 1694–98.

Cirio, M.C., Martel, J., Mann, M., Toppings, M., Bartolomei, M., Trasler, J., and Chaillet, J.R. (2008a). DNA methyltransferase 1o functions during preimplantation development to preclude a profound level of epigenetic variation. *Dev Biol* 324, pp. 139–50.

Cirio, M.C., Ratnam, S., Ding, F., Reinhart, B., Navara, C., and Chaillet, J.R. (2008b). Preimplantation expression of the somatic form of Dnmt1 suggests a role in the inheritance of genomic imprints. *BMC Dev Biol* 8, p. 9.

Ciruna, B., Jenny, A., Lee, D., Mlodzik, M., and Schier, A.F. (2006). Planar cell polarity signalling couples cell division and morphogenesis during neurulation. *Nature* 439, pp. 220–24.

Ciruna, B., Weidinger, G., Knaut, H., Thisse, B., Thisse, C., Raz, E., and Schier, A.F. (2002). Production of maternal-zygotic mutant zebrafish by germline replacement. *Proc Natl Acad Sci U S A* 99, pp. 14919–24.

Claussen, M., and Pieler, T. (2004). Xvelo1 uses a novel 75-nucleotide signal sequence that drives vegetal localization along the late pathway in *Xenopus* oocytes. *Dev Biol* 266, pp. 270–84.

Clegg, K.B., and Piko, L. (1982). RNA synthesis and cytoplasmic polyadenylation in the one-cell mouse embryo. *Nature* 295, pp. 343–44.

Clelland, E., and Peng, C. (2009). Endocrine/paracrine control of zebrafish ovarian development. *Mol Cell Endocrinol* 312, pp. 42–52.

Colledge, W.H., Carlton, M.B., Udy, G.B., and Evans, M.J. (1994). Disruption of c-mos causes parthenogenetic development of unfertilized mouse eggs. *Nature* 370, pp. 65–8.

Collier, B., Gorgoni, B., Loveridge, C., Cooke, H.J., and Gray, N.K. (2005). The DAZL family proteins are PABP-binding proteins that regulate translation in germ cells. *Embo J* 24, pp. 2656–66.

Colombo, K., Grill, S.W., Kimple, R.J., Willard, F.S., Siderovski, D.P., and Gonczy, P. (2003). Translation of polarity cues into asymmetric spindle positioning in *Caenorhabditis elegans* embryos. *Science* 300, pp. 1957–61.

Colonna, R., Cecconi, S., Tatone, C., Mangia, F., and Buccione, R. (1989). Somatic cell-oocyte interactions in mouse oogenesis: stage-specific regulation of mouse oocyte protein phosphorylation by granulosa cells. *Dev Biol* 133, pp. 305–8.

Constam, D.B., and Robertson, E.J. (2000). SPC4/PACE4 regulates a TGFbeta signaling network during axis formation. *Genes Dev* 14, pp. 1146–55.

Coonrod, S.A., Bolling, L.C., Wright, P.W., Visconti, P.E., and Herr, J.C. (2001). A morpholino phenocopy of the mouse mos mutation. *Genesis* 30, pp. 198–200.

Cote, C.A., Gautreau, D., Denegre, J.M., Kress, T.L., Terry, N.A., and Mowry, K.L. (1999). A *Xenopus* protein related to hnRNP I has a role in cytoplasmic RNA localization. *Mol Cell* 4, pp. 431–7.

Cox, D.N., Chao, A., Baker, J., Chang, L., Qiao, D., and Lin, H. (1998). A novel class of evolutionarily conserved genes defined by piwi are essential for stem cell self-renewal. *Genes Dev* 12, pp. 3715–27.

Cox, D.N., Chao, A., and Lin, H. (2000). piwi encodes a nucleoplasmic factor whose activity modulates the number and division rate of germline stem cells. *Development* 127, pp. 503–14.

Cox, D.N., Lu, B., Sun, T.Q., Williams, L.T., and Jan, Y.N. (2001). Drosophila par-1 is required for oocyte differentiation and microtubule organization. *Curr Biol* 11, pp. 75–87.

Cox, R.T., and Spradling, A.C. (2003). A Balbiani body and the fusome mediate mitochondrial inheritance during *Drosophila* oogenesis. *Development* 130, pp. 1579–90.

Cox, R.T., and Spradling, A.C. (2006). Milton controls the early acquisition of mitochondria by *Drosophila* oocytes. *Development* 133, pp. 3371–77.

Cui, X.S., Li, X.Y., and Kim, N.H. (2007). Cdc42 is implicated in polarity during meiotic resumption and blastocyst formation in the mouse. *Mol Reprod Dev* 74, pp. 785–94.

Cui, Y., Brown, J.D., Moon, R.T., and Christian, J.L. (1995). Xwnt-8b: a maternally expressed *Xenopus* Wnt gene with a potential role in establishing the dorsoventral axis. *Development* 121, pp. 2177–86.

Cuykendall, T.N., and Houston, D.W. (2009). Vegetally localized *Xenopus* trim36 regulates cortical rotation and dorsal axis formation. *Development* 136, pp. 3057–65.

Czaplinski, K., and Mattaj, I.W. (2006). 40LoVe interacts with Vg1 RBP/ Vera and hnRNP I in binding the Vg1-localization element. *RNA* 12, pp. 213–22.

Danilchik, M.V., Funk, W.C., Brown, E.E., and Larkin, K. (1998). Requirement for microtubules in new membrane formation during cytokinesis of *Xenopus* embryos. *Dev Biol* 194, pp. 47–60.

Davis, A.C., Wims, M., Spotts, G.D., Hann, S.R., and Bradley, A. (1993). A null c myc mutation causes lethality before 10.5 days of gestation in homozygotes and reduced fertility in heterozygous female mice. *Genes Dev* 7, pp. 671–82.

Davis, W., Jr., De Sousa, P.A., and Schultz, R.M. (1996). Transient expression of translation initiation factor eIF-4C during the 2-cell stage of the preimplantation mouse embryo: identification by mRNA differential display and the role of DNA replication in zygotic gene activation. *Dev Biol* 174, pp. 190–201.

Davis, T.L., Yang, G.J., McCarrey, J.R., and Bartolomei, M.S. (2000). The H19 methylation imprint is erased and re-established differentially on the parental alleles during male germ cell development. *Hum Mol Genet* 9, pp. 2885–94.

De, J., Lai, W.S., Thorn, J.M., Goldsworthy, S.M., Liu, X., Blackwell, T.K., and Blackshear, P.J. (1999). Identification of four CCCH zinc finger proteins in *Xenopus*, including a novel vertebrate protein with four zinc fingers and severely restricted expression. *Gene* 228, pp. 133–45.

De Cuevas, M., and Spradling, A.C. (1998). Morphogenesis of the *Drosophila* fusome and its implications for oocyte specification. *Development* 125, pp. 2781–9.

De Felici, M. (2001). Twenty years of research on primordial germ cells. *Int J Dev Biol* 45, pp. 519–22.

De La Fuente, R., and Eppig, J.J. (2001). Transcriptional activity of the mouse oocyte genome: companion granulosa cells modulate transcription and chromatin remodeling. *Dev Biol* 229, pp. 224–36.

De La Fuente, R., Viveiros, M.M., Burns, K.H., Adashi, E.Y., Matzuk, M.M., and Eppig, J.J. (2004). Major chromatin remodeling in the germinal vesicle (GV) of mammalian oocytes is dispensable for global transcriptional silencing but required for centromeric heterochromatin function. *Dev Biol* 275, pp. 447–58.

de Smedt, V., Szollosi, D., and Kloc, M. (2000). The balbiani body: asymmetry in the mammalian oocyte. *Genesis* 26, pp. 208–12.

De Vries, W.N., Evsikov, A.V., Haac, B.E., Fancher, K.S., Holbrook, A.E., Kemler, R., Solter, D., and Knowles, B.B. (2004). Maternal beta-catenin and E cadherin in mouse development. *Development* 131, pp. 4435–45.

Debey, P., Szollosi, M.S., Szollosi, D., Vautier, D., Girousse, A., and Besombes, D. (1993). Competent mouse oocytes isolated from antral follicles exhibit different chromatin organization and follow different maturation dynamics. *Mol Reprod Dev* 36, pp. 59–74.

Dekens, M.P., Pelegri, F.J., Maischein, H.M., and Nusslein-Volhard, C. (2003). The maternal-effect gene futile cycle is essential for pronuclear congression and mitotic spindle assembly in the zebrafish zygote. *Development* 130, pp. 3907–16.

Deng, W., and Lin, H. (1997). Spectrosomes and fusomes anchor mitotic spindles during asymmetric germ cell divisions and facilitate the formation of a polarized microtubule array for oocyte specification in *Drosophila*. *Dev Biol* 189, pp. 79–94.

Deng, W., and Lin, H. (2001). Asymmetric germ cell division and oocyte determination during *Drosophila* oogenesis. *Int Rev Cytol* 203, pp. 93–138.

Deng, M., Suraneni, P., Schultz, R.M., and Li, R. (2007). The Ran GTPase mediates chromatin signaling to control cortical polarity during polar body extrusion in mouse oocytes. *Dev Cell* 12, pp. 301–08.

Deng, M., Williams, C.J., and Schultz, R.M. (2005). Role of MAP kinase and myosin light chain kinase in chromosome-induced development of mouse egg polarity. *Dev Biol* 278, pp. 358–66.

Deng, W., and Lin, H. (2002). Miwi, a murine homolog of piwi, encodes a cytoplasmic protein essential for spermatogenesis. *Dev Cell* 2, pp. 819–30.

Diaz, F.J., Wigglesworth, K., and Eppig, J.J. (2007). Oocytes determine cumulus cell lineage in mouse ovarian follicles. *J Cell Sci* 120, pp. 1330–40.

Dierich, A., Sairam, M.R., Monaco, L., Fimia, G.M., Gansmuller, A., LeMeur, M., and Sassone-Corsi, P. (1998). Impairing follicle-stimulating hormone (FSH) signaling in vivo: targeted disruption of the FSH receptor leads to aberrant gametogenesis and hormonal imbalance. *Proc Natl Acad Sci U S A* 95, pp. 13612–17.

Dilworth, S.M., Black, S.J., and Laskey, R.A. (1987). Two complexes that contain histones are required for nucleosome assembly in vitro: role of nucleoplasmin and N1 in *Xenopus* egg extracts. *Cell* 51, pp. 1009–18.

Dingwall, C., Dilworth, S.M., Black, S.J., Kearsey, S.E., Cox, L.S., and Laskey, R.A. (1987). Nucleoplasmin cDNA sequence reveals polyglutamic acid tracts and a cluster of sequences homologous to putative nuclear localization signals. *EMBO J* 6, pp. 69–74.

Djagaeva, I., Doronkin, S., and Beckendorf, S.K. (2005). Src64 is involved in fusome development and karyosome formation during *Drosophila* oogenesis. *Dev Biol* 284, pp. 143–56.

Djiane, A., Riou, J., Umbhauer, M., Boucaut, J., and Shi, D. (2000). Role of frizzled 7 in the regulation of convergent extension movements during gastrulation in *Xenopus laevis*. *Development* 127, pp. 3091–100.

Dodson, G.S., Guarnieri, D.J., and Simon, M.A. (1998). Src64 is required for ovarian ring canal morphogenesis during *Drosophila* oogenesis. *Development* 125, pp. 2883–92.

Dong, M., Fu, Y.F., Du, T.T., Jing, C.B., Fu, C.T., Chen, Y., Jin, Y., Deng, M., and Liu, T.X. (2009). Heritable and lineage-specific gene knockdown in zebrafish embryo. *PLoS One* 4, p. e6125.

Dorsky, R.I., Itoh, M., Moon, R.T., and Chitnis, A. (2003). Two tcf3 genes cooperate to pattern the zebrafish brain. *Development* 130, pp. 1937–47.

Dorsky, R.I., Snyder, A., Cretekos, C.J., Grunwald, D.J., Geisler, R., Haffter, P., Moon, R.T., and Raible, D.W. (1999). Maternal and embryonic expression of zebrafish lef1. *Mech Dev* 86, pp. 147–50.

Dosch, R., Wagner, D.S., Mintzer, K.A., Runke, G., Wiemelt, A.P., and Mullins, M.C. (2004). Maternal control of vertebrate development before the midblastula transition: mutants from the zebrafish I. *Dev Cell* 6, pp. 771–80.

Dovey, M., Patton, E.E., Bowman, T., North, T., Goessling, W., Zhou, Y., and Zon, L.I. (2009). Topoisomerase II alpha is required for embryonic development and liver regeneration in zebrafish. *Mol Cell Biol* 29, pp. 3746–53.

Draetta, G., and Beach, D. (1989). The mammalian cdc2 protein kinase: mechanisms of regulation during the cell cycle. *J Cell Sci* Suppl 12, pp. 21–7.

Draper, B.W., McCallum, C.M., and Moens, C.B. (2007). nanos1 is required to maintain oocyte production in adult zebrafish. *Dev Biol* 305, pp. 589–98.

Draper, B.W., Mello, C.C., Bowerman, B., Hardin, J., and Priess, J.R. (1996). MEX-3 is a KH domain protein that regulates blastomere identity in early *C. elegans* embryos. *Cell* 87, pp. 205–16.

Driever, W., Solnica-Krezel, L., Schier, A.F., Neuhauss, S.C., Malicki, J., Stemple, D.L., Stainier, D.Y., Zwartkruis, F., Abdelilah, S., Rangini, Z., et al. (1996). A genetic screen for mutations affecting embryogenesis in zebrafish. *Development* 123, pp. 37–46.

Dumont, J., Million, K., Sunderland, K., Rassinier, P., Lim, H., Leader, B., and Verlhac, M.H. (2007). Formin-2 is required for spindle migration and for the late steps of cytokinesis in mouse oocytes. *Dev Biol* 301, pp. 254–65.

Duncan, F.E., Moss, S.B., Schultz, R.M., and Williams, C.J. (2005). PAR-3 defines a central subdomain of the cortical actin cap in mouse eggs. *Dev Biol* 280, pp. 38–47.

Dunican, D.S., Ruzov, A., Hackett, J.A., and Meehan, R.R. (2008). xDnmt1 regulates transcriptional silencing in pre-MBT *Xenopus* embryos independently of its catalytic function. *Development* 135, pp. 1295–1302.

Dworkin, M.B., and Dworkin-Rastl, E. (1990). Functions of maternal mRNA in early development. *Mol Reprod Dev* 26, pp. 261–97.

Edelmann, W., Cohen, P.E., Kane, M., Lau, K., Morrow, B., Bennett, S., Umar, A., Kunkel, T., Cattoretti, G., Chaganti, R., et al. (1996). Meiotic pachytene arrest in MLH1-deficient mice. *Cell* 85, pp. 1125–34.

Edwards, C.A., and Ferguson-Smith, A.C. (2007). Mechanisms regulating imprinted genes in clusters. *Curr Opin Cell Biol* 19, pp. 281–89.

Elinson, R.P. (1975). Site of sperm entry and a cortical contraction associated with egg activation in the frog Rana pipiens. *Dev Biol* 47, pp. 257–68.

Elinson, R.P., and Rowning, B. (1988). A transient array of parallel microtubules in frog eggs: Potential tracks for a cytoplasmic rotation that specifies the dorso ventral axis. *Dev Biol* 128, pp. 185–97.

Ephrussi, A., and Lehmann, R. (1992). Induction of germ cell formation by *oskar*. *Nature* 358, pp. 387–91.

Eppig, J.J. (1996). Coordination of nuclear and cytoplasmic oocyte maturation in eutherian mammals. *Reprod Fertil Dev* 8, pp. 485–89.

Eppig, J.J. (2001). Oocyte control of ovarian follicular development and function in mammals. *Reproduction* 122, pp. 829–38.

Erter, C.E., Wilm, T.P., Basler, N., Wright, C.V., and Solnica-Krezel, L. (2001). Wnt8 is required in lateral mesendodermal precursors for neural posteriorization in vivo. *Development* 128, pp. 3571–83.

Esposito, G., Vitale, A.M., Leijten, F.P., Strik, A.M., Koonen-Reemst, A.M., Yurttas, P., Robben, T.J., Coonrod, S., and Gossen, J.A. (2007). Peptidylarginine deiminase (PAD) 6 is essential for oocyte cytoskeletal sheet formation and female fertility. *Mol Cell Endocrinol* 273, pp. 25–31.

Ewen-Campen, B., Schwager, E.E., and Extavour, C.G. The molecular machinery of germ line specification. *Mol Reprod Dev* 77, pp. 3–18.

Fanto, M., and McNeill, H. (2004). Planar polarity from flies to vertebrates. *J Cell Sci* 117, pp. 527–33.

Fedoriw, A.M., Stein, P., Svoboda, P., Schultz, R.M., and Bartolomei, M.S. (2004). Transgenic RNAi reveals essential function for CTCF in H19 gene imprinting. *Science* 303, pp. 238–40.

Feitsma, H., Leal, M.C., Moens, P.B., Cuppen, E., and Schulz, R.W. (2007). Mlh1 deficiency in zebrafish results in male sterility and aneuploid as well as triploid progeny in females. *Genetics* 175, pp. 1561–69.

Fekany-Lee, K., Gonzalez, E., Miller-Bertoglio, V., and Solnica-Krezel, L. (2000). The homeobox gene *bozozok* promotes anterior neuroectoderm formation in zebrafish through negative regulation of BMP2/4 and Wnt pathways. *Development* 127, pp. 2333–45.

Feng, B., Schwarz, H., and Jesuthasan, S. (2002). Furrow-specific endocytosis during cytokinesis of zebrafish blastomeres. *Exp Cell Res* 279, pp. 14–20.

Ferg, M., Sanges, R., Gehrig, J., Kiss, J., Bauer, M., Lovas, A., Szabo, M., Yang, L., Straehle, U., Pankratz, M.J., et al. (2007). The TATA-binding protein regulates maternal mRNA degradation and differential zygotic transcription in zebrafish. *Embo J* 26, pp. 3945–56.

Fernandez, J., Valladares, M., Fuentes, R., and Ubilla, A. (2006). Reorganization of cytoplasm in the zebrafish oocyte and egg during early steps of ooplasmic segregation. *Dev Dyn* 235, pp. 656–71.

Ferrell, J.E., Jr. (1999). *Xenopus* oocyte maturation: new lessons from a good egg. *Bioessays* 21, pp. 833–42.

Ferrell, J.E., Jr., Wu, M., Gerhart, J.C., and Martin, G.S. (1991). Cell cycle tyrosine phosphorylation of p34cdc2 and a microtubule-associated protein kinase homolog in *Xenopus* oocytes and eggs. *Mol Cell Biol* 11, pp. 1965–71.

Ferrell, J.E., Jr., Pomerening, J.R., Kim, S.Y., Trunnell, N.B., Xiong, W., Huang, C.Y., and Machleder, E.M. (2009). Simple, realistic models of complex biological processes: positive feedback and bistability in a cell fate switch and a cell cycle oscillator. *FEBS Lett* 583, pp. 3999–4005.

Filosa, S., and Taddei, C. (1976). Intercellular bridges in lizard oogenesis. *Cell Differ* 5, pp. 199–206.

Fire, A., Xu, S., Montgomery, M.K., Kostas, S.A., Driver, S.E., and Mello, C.C. (1998). Potent and specific genetic interference by double-stranded RNA in *Caenorhabditis elegans*. *Nature* 391, pp. 806–11.

Fitzpatrick, G.V., Pugacheva, E.M., Shin, J.Y., Abdullaev, Z., Yang, Y., Khatod, K., Lobanenkov, V.V., and Higgins, M.J. (2007). Allele-specific binding of CTCF to the multipartite imprinting control region KvDMR1. *Mol Cell Biol* 27, pp. 2636–47.

Flores, M.V., Lam, E.Y., Crosier, K.E., and Crosier, P.S. (2008). Osteogenic transcription factor Runx2 is a maternal determinant of dorsoventral patterning in zebrafish. *Nat Cell Biol* 10, pp. 346–52.

Fojas de Borja, P., Collins, N.K., Du, P., Azizkhan-Clifford, J., and Mudryj, M. (2001). Cyclin A-CDK phosphorylates Sp1 and enhances Sp1-mediated transcription. *Embo J* 20, pp. 5737–47.

Foygel, K., Choi, B., Jun, S., Leong, D.E., Lee, A., Wong, C.C., Zuo, E., Eckart, M., Reijo Pera, R.A., Wong, W.H., et al. (2008). A novel and critical role for Oct4 as a regulator of the maternal-embryonic transition. *PLoS One* 3, p. e4109.

Franco, H.L., Lee, K.Y., Broaddus, R.R., White, L.D., Lanske, B., Lydon, J.P., Jeong, J.W., and DeMayo, F.J. (2010a). Ablation of Indian hedgehog in the murine uterus results in decreased cell cycle progression, aberrant epidermal growth factor signaling, and increased estrogen signaling. *Biol Reprod* 82, pp. 783–90.

Franco, H.L., Lee, K.Y., Rubel, C.A., Creighton, C.J., White, L.D., Broaddus, R.R., Lewis, M.T., Lydon, J.P., Jeong, J.W., and DeMayo, F.J. (2010b). Constitutive activation of smoothened leads to female infertility and altered uterine differentiation in the mouse. *Biol Reprod* 82, pp. 991–9.

Fuchimoto, D., Mizukoshi, A., Schultz, R.M., Sakai, S., and Aoki, F. (2001). Posttranscriptional regulation of cyclin A1 and cyclin A2 during mouse oocyte meiotic maturation and preimplantation development. *Biol Reprod* 65, pp. 986–93.

Fujisue, M., Kobayakawa, Y., and Yamana, K. (1993). Occurrence of dorsal axisinducing activity around the vegetal pole of an uncleaved *Xenopus* egg and displacement to the equatorial region by cortical rotation. *Development* 118, pp. 163–70.

Fukazawa, C., Santiago, C., Park, K.M., Deery, W.J., Gomez de la Torre, Canny, S., Holterhoff, C.K., and Wagner, D.S. (2010). poky/chuk/ikk1 is required for differentiation of the zebrafish embryonic epidermis. *Dev Biol* 346, pp. 272–83.

Fuller, M.T. (1998). Genetic control of cell proliferation and differentiation in *Drosophila* spermatogenesis. *Semin Cell Dev Biol* 9, pp. 433–4.

Funayama, N., Fagotto, F., McCrea, P., and Gumbiner, B.M. (1995). Embryonic axis induction by the armadillo repeat domain of b-catenin: evidence for intracellular signaling. *J Cell Biol* 128, pp. 959–68.

Gallicano, G.I., Larabell, C.A., McGaughey, R.W., and Capco, D.G. (1994a). Novel cytoskeletal elements in mammalian eggs are composed of a unique arrangement of intermediate filaments. *Mech Dev* 45, pp. 211–26.

Gallicano, G.I., McGaughey, R.W., and Capco, D.G. (1994b). Ontogeny of the cytoskeleton during mammalian oogenesis. *Microsc Res Tech* 27, pp. 134–44.

Gardner, A.J., and Evans, J.P. (2006). Mammalian membrane block to polyspermy: new insights into how mammalian eggs prevent fertilisation by multiple sperm. *Reprod Fertil Dev* 18, pp. 53–61.

Gardner, R.L. (2007). The axis of polarity of the mouse blastocyst is specified before blastulation and independently of the zona pellucida. *Hum Reprod* 22, pp. 798–806.

Gautier, J., Minshull, J., Lohka, M., Glotzer, M., Hunt, T., and Maller, J.L. (1990). Cyclin is a component of maturation-promoting factor from *Xenopus*. *Cell* 60, pp. 487–94.

Gerhart, J., Danilchik, M., Doniach, T., Roberts, S., Rowning, B., and Stewart, R. (1989). Cortical rotation of the *Xenopus* egg: consequences for the anteroposterior pattern of embryonic dorsal development. *Development* 1989 (Suppl), pp. 37–51.

Gibert, Y., Gajewski, A., Meyer, A., and Begemann, G. (2006). Induction and prepatterning of the zebrafish pectoral fin bud requires axial retinoic acid signaling. *Development* 133, pp. 2649–59.

Gilboa, L., and Lehmann, R. (2004). Repression of primordial germ cell differentiation parallels germ line stem cell maintenance. *Curr Biol* 14, pp. 981–6.

Gillespie, P.J., and Blow, J.J. (2000). Nucleoplasmin-mediated chromatin remodelling is required for *Xenopus* sperm nuclei to become licensed for DNA replication. *Nucleic Acids Res* 28, pp. 472–80.

Giraldez, A.J., Cinalli, R.M., Glasner, M.E., Enright, A.J., Thomson, J.M., Baskerville, S., Hammond, S.M., Bartel, D.P., and Schier, A.F. (2005). MicroRNAs regulate brain morphogenesis in zebrafish. *Science* 308, pp. 833–8.

Giraldez, A.J., Mishima, Y., Rihel, J., Grocock, R.J., Van Dongen, S., Inoue, K., Enright, A.J., and Schier, A.F. (2006). Zebrafish MiR-430 promotes deadenylation and clearance of maternal mRNAs. *Science* 312, pp. 75–9.

Goldstein, B., and Macara, I.G. (2007). The PAR proteins: fundamental players in animal cell polarization. *Dev Cell* 13, pp. 609–22.

Gonczy, P., Matunis, E., and DiNardo, S. (1997). Bag-of-marbles and benign gonial cell neoplasm act in the germline to restrict proliferation during Drosophila spermatogenesis. *Development* 124, pp. 4361–71.

Gondos, B. (1987). Comparative studies of normal and neoplastic ovarian germ cells: 2. Ultrastructure and pathogenesis of dysgerminoma. *Int J Gynecol Pathol* 6, pp. 124–31.

Gonzalez, M.A., Tachibana, K.E., Adams, D.J., van der Weyden, L., Hemberger, M., Coleman, N., Bradley, A., and Laskey, R.A. (2006). Geminin is essential to prevent endoreduplication and to form pluripotent cells during mammalian development. *Genes Dev* 20, pp. 1880–4.

Gore, A.V., Maegawa, S., Cheong, A., Gilligan, P.C., Weinberg, E.S., and Sampath, K. (2005). The zebrafish dorsal axis is apparent at the four-cell stage. *Nature* 438, pp. 1030–5.

Gorjanacz, M., Adam, G., Torok, I., Mechler, B.M., Szlanka, T., and Kiss, I. (2002). Importin-alpha 2 is critically required for the assembly of ring canals during *Drosophila* oogenesis. *Dev Biol* 251, pp. 271–82.

Gorjanacz, M., Torok, I., Pomozi, I., Garab, G., Szlanka, T., Kiss, I., and Mechler, B.M. (2006). Domains of Importin-alpha2 required for ring canal assembly during *Drosophila* oogenesis. *J Struct Biol* 154, pp. 27–41.

Goutel, C., Kishimoto, Y., Schulte-Merker, S., and Rosa, F. (2000). The ventralizing activity of Radar, a maternally expressed bone morphogenetic protein, reveals complex bone morphogenetic protein interactions controlling dorsoventral patterning in zebrafish. *Mech Dev* 99, pp. 15–27.

Greenbaum, M.P., Iwamori, N., Agno, J.E., and Matzuk, M.M. (2009). Mouse TEX14 is required for embryonic germ cell intercellular bridges but not female fertility. *Biol Reprod* 80, pp. 449–57.

Grieder, N.C., de Cuevas, M., and Spradling, A.C. (2000). The fusome organizes the microtubule network during oocyte differentiation in *Drosophila*. *Development* 127, pp. 4253–64.

Griffin, K.J., Amacher, S.L., Kimmel, C.B., and Kimelman, D. (1998). Molecular identification of spadetail: regulation of zebrafish trunk and tail mesoderm formation by T-box genes. *Development* 125, pp. 3379–88.

Grishok, A., Pasquinelli, A.E., Conte, D., Li, N., Parrish, S., Ha, I., Baillie, D.L., Fire, A., Ruvkun, G., and Mello, C.C. (2001). Genes and mechanisms related to RNA interference regulate expression of the small temporal RNAs that control *C. elegans* developmental timing. *Cell* 106, pp. 23–34.

Gritsman, K., Zhang, J., Cheng, S., Heckscher, E., Talbot, W.S., and Schier, A.F. (1999). The EGF-CFC protein one-eyed pinhead is essential for nodal signaling. *Cell* 97, pp. 121–32.

Grivna, S.T., Pyhtila, B., and Lin, H. (2006). MIWI associates with translational machinery and PIWI-interacting RNAs (piRNAs) in regulating spermatogenesis. *Proc Natl Acad Sci U S A* 103, pp. 13415–20.

Guger, K.A., and Gumbiner, B.M. (1995). β-catenin has Wnt-like activity and mimics the Nieuwkoop signaling center in *Xenopus* dorsal–ventral patterning. *Dev Biol* 172, pp. 115–25.

Guo, X., and Gao, S. (2009). Pins homolog LGN regulates meiotic spindle organization in mouse oocytes. *Cell Res* 19, pp. 838–48.

Gupta, T., Marlow, F.L., Ferriola, D., Mackiewicz, K., Dapprich, J., Monos, D., and Mullins, M.C. (2010). Microtubule actin crosslinking factor 1 regulates balbiani body function and animal–vegetal polarity of the zebrafish oocyte. *PLoS Genet* 6 (8), p. e1001073.

Guraya, S.S. (1979). Recent advances in the morphology, cytochemistry, and function of Balbiani's vitelline body in animal oocytes. *Int Rev Cytol* 59, pp. 249–321.

Gurtu, V.E., Verma, S., Grossmann, A.H., Liskay, R.M., Skarnes, W.C., and Baker, S.M. (2002). Maternal effect for DNA mismatch repair in the mouse. *Genetics* 160, pp. 271–77.

Haccard, O., Jessus, C., Cayla, X., Goris, J., Merlevede, W., and Ozon, R. (1990). In vivo activation of a microtubule-associated protein kinase during meiotic maturation of the *Xenopus* oocyte. *Eur J Biochem* 192, pp. 633–42.

Hagos, E.G., Fan, X., and Dougan, S.T. (2007). The role of maternal Activin-like signals in zebrafish embryos. *Dev Biol* 309, pp. 245–58.

Halet, G., and Carroll, J. (2007). Rac activity is polarized and regulates meiotic spindle stability and anchoring in mammalian oocytes. *Dev Cell* 12, pp. 309–17.

Hamade, A., Deries, M., Begemann, G., Bally-Cuif, L., Genet, C., Sabatier, F., Bonnieu, A., and Cousin, X. (2006). Retinoic acid activates myogenesis in vivo through Fgf8 signalling. *Dev Biol* 289, pp. 127–40.

Hamatani, T., Carter, M.G., Sharov, A.A., and Ko, M.S. (2004). Dynamics of global gene expression changes during mouse preimplantation development. *Dev Cell* 6, pp. 117–31.

Hammond, S.M., Bernstein, E., Beach, D., and Hannon, G.J. (2000). An RNAdirected nuclease mediates post-transcriptional gene silencing in *Drosophila* cells. *Nature* 404, pp. 293–6.

Hammond, S.S., and Matin, A. (2009). Tools for the genetic analysis of germ cells. *Genesis* 47, pp. 617–27.

Hara, K.T., Oda, S., Naito, K., Nagata, M., Schultz, R.M., and Aoki, F. (2005). Cyclin A2-CDK2 regulates embryonic gene activation in 1-cell mouse embryos. *Dev Biol* 286, pp. 102–13.

Hara, K., Nakayama, K.I., and Nakayama, K. (2006). Geminin is essential for the development of preimplantation mouse embryos. *Genes Cells* 11, pp. 1281–93.

Hart, N.H. (1990). Fertilization in teleost fishes: mechanisms of sperm-egg interactions. *Int Rev* Cytol 121, pp. 1–66.

Hart, N.H., and Donovan, M. (1983). Fine structure of the chorion and site of sperm entry in the egg of *Brachydanio*. *J Exp Zool* 227, pp. 277–96.

Hart, N.H., and Fluck, R.A. (1995). Cytoskeleton in teleost eggs and early embryos: contributions to cytoarchitecture and motile events. *Curr Top Dev Biol* 31, pp. 343–81.

Hashimoto, Y., Maegawa, S., Nagai, T., Yamaha, E., Suzuki, H., Yasuda, K., and Inoue, K. (2004). Localized maternal factors are required for zebrafish germ cell formation. *Dev Biol* 268, pp. 152–61.

Hatada, S., Kinoshita, M., Takahashi, S., Nishihara, R., Sakumoto, H., Fukui, A., Noda, M., and Asashima, M. (1997). An interferon regulatory factor-related gene (xIRF-6) is expressed in the posterior mesoderm during the early development of *Xenopus laevis*. *Gene* 203, pp. 183–8.

Hayashi, S., Tenzen, T., and McMahon, A.P. (2003). Maternal inheritance of Cre activity in a Sox-2Cre deleter strain. *Genesis* 37, pp. 51–53.

Heasman, J., Ginsberg, D., Geiger, B., Goldstone, K., Pratt, T., Yoshida-Noro, C., and Wylie, C. (1994). A functional test for maternally inherited cadherin in *Xenopus* shows its importance in cell adhesion at the blastula stage. *Development* 120, pp. 49–57.

Heasman, J., Wessely, O., Langland, R., Craig, E.J., and Kessler, D.S. (2001). Vegetal localization of maternal mRNAs is disrupted by VegT depletion. *Dev Biol* 240, pp. 377–86.

Heisenberg, C.P., Houart, C., Take-Uchi, M., Rauch, G.J., Young, N., Coutinho, P., Masai, I., Caneparo, L., Concha, M.L., Geisler, R., et al. (2001). A mutation in the Gsk3-binding domain of zebrafish Masterblind/Axin1 leads to a fate transformation of telencephalon and eyes to diencephalon. *Genes Dev* 15, pp. 1427–34.

Hedgepeth, C.M., Conrad, L.J., Zhang, J., Huang, H.C., Lee, V.M., and Klein, P.S. (1997). Activation of the Wnt signaling pathway: a molecular mechanism for lithium action. *Dev Biol* 185, pp. 82–91.

Heisenberg, C.P., Tada, M., Rauch, G.J., Saude, L., Concha, M.L., Geisler, R., Stemple, D.L., Smith, J.C.F., and Wilson, S.W. (2000). Silberblick / Wnt11 mediates convergent extension movements during zebrafish gastrulation. *Nature* 405, pp. 76–81.

Helde, K.A., and Grunwald, D.J. (1993). The DVR-1 (Vg1) transcript of zebrafish is maternally supplied and distributed throughout the embryo. *Dev Biol* 159, pp. 418–26.

Henery, C.C., Miranda, M., Wiekowski, M., Wilmut, I., and DePamphilis, M.L. (1995). Repression of gene expression at the beginning of mouse development. *Dev Biol* 169, pp. 448–60.

Henry, C.A., McNulty, I.M., Durst, W.A., Munchel, S.E., and Amacher, S.L. (2005). Interactions between muscle fibers and segment boundaries in zebrafish. *Dev Biol* 287, pp. 346–60.

Hernandez-Gonzalez, I., Gonzalez-Robayna, I., Shimada, M., Wayne, C.M., Ochsner, S.A., White, L., and Richards, J.S. (2006). Gene expression profiles of cumulus cell oocyte complexes during ovulation reveal cumulus cells express neuronal and immune-related genes: does this expand their role in the ovulation process? *Mol Endocrinol* 20, pp. 1300–21.

Herr, J.C., Chertihin, O., Digilio, L., Jha, K.N., Vemuganti, S., and Flickinger, C.J. (2008). Distribution of RNA binding protein MOEP19 in the oocyte cortex and early embryo indicates pre-patterning related to blastomere polarity and trophectoderm specification. *Dev Biol* 314, pp. 300–16.

Hibi, M., Hirano, T., and Dawid, I.B. (2002). Organizer formation and function. *Results Probl Cell Differ* 40, pp. 48–71.

Hikichi, T., Kohda, T., Kaneko-Ishino, T., and Ishino, F. (2003). Imprinting regulation of the murine Meg1/Grb10 and human GRB10 genes; roles of brainspecific promoters and mouse-specific CTCF-binding sites. *Nucleic Acids Res* 31, pp. 1398–406.

Hild, M., Dick, A., Rauch, G.J., Meier, A., Bouwmeester, T., Haffter, P., and Hammerschmidt, M. (1999). The smad5 mutation somitabun blocks Bmp2b signaling during early dorsoventral patterning of the zebrafish embryo. *Development* 126, pp. 2149–59.

Hirasawa, R., Chiba, H., Kaneda, M., Tajima, S., Li, E., Jaenisch, R., and Sasaki, H. (2008). Maternal and zygotic Dnmt1 are necessary and sufficient for the maintenance of DNA methylation imprints during preimplantation development. *Genes Dev* 22, pp. 1607–16.

Hirasawa, R., and Feil, R. (2008). A KRAB domain zinc finger protein in imprinting and disease. *Dev Cell* 15, pp. 487–8.

Hirasawa, R., and Sasaki, H. (2009). Dynamic transition of Dnmt3b expression in mouse pre- and early post-implantation embryos. *Gene Expr Patterns* 9, pp. 27–30.

Ho, R.K. (1992). Cell movements and cell fate during zebrafish gastrulation. *Dev Suppl*, pp. 65–73.

Holloway, B.A., Gomez de la Torre Canny, S., Ye, Y., Slusarski, D.C., Freisinger, C.M., Dosch, R., Chou, M.M., Wagner, D.S., and Mullins, M.C. (2009). A novel role for MAPKAPK2 in morphogenesis during zebrafish development. *PLoS Genet* 5, p. e1000413.

Holowacz, T., and Elinson, R.P. (1993). Cortical cytoplasm, which induces dorsal axis formation in *Xenopus*, is inactivate by UV irradiation of the oocyte. *Development* 119, pp. 277–85.

Hoodless, P.A., Pye, M., Chazaud, C., Labbe, E., Attisano, L., Rossant, J., and Wrana, J.L. (2001). FoxH1 (Fast) functions to specify the anterior primitive streak in the mouse. *Genes Dev* 15, pp. 1257–71.

Houliston, E., Le Guellec, R., Kress, M., Philippe, M., and Le Guellec, K. (1994). The kinesin-related protein Eg5 associates with both interphase and spindle microtubules during *Xenopus* early development. *Dev Biol* 164, pp. 147–59.

Houston, D.W., Zhang, J., Maines, J.Z., Wasserman, S.A., and King, M.L. (1998). A *Xenopus* DAZ-like gene encodes an RNA component of germ plasm and is a functional homologue of *Drosophila* boule. *Development* 125, pp. 171–80.

Houston, D.W., and King, M.L. (2000). A critical role for Xdazl, a germ plasm-localized RNA, in the differentiation of primordial germ cells in *Xenopus*. *Development* 127, pp. 447–56.

Houston, D.W., and Wylie, C. (2002). Cloning and expression of *Xenopus* Lrp5 and Lrp6 genes. *Mech Dev* 117, pp. 337–42.

Houston, D.W., and Wylie, C. (2005). Maternal *Xenopus* Zic2 negatively regulates Nodal-related gene expression during anteroposterior patterning. *Development* 132, pp. 4845–55.

Houston, D.W., Kofron, M., Resnik, E., Langland, R., Destree, O., Wylie, C., and Heasman, J. (2002). Repression of organizer genes in dorsal and ventral Xenopus cells mediated by maternal XTcf3. *Development* 129, pp. 4015–25.

Houwing, S., Berezikov, E., and Ketting, R.F. (2008). Zili is required for germ cell differentiation and meiosis in zebrafish. *EMBO J* 27, pp. 2702–11.

Houwing, S., Kamminga, L.M., Berezikov, E., Cronembold, D., Girard, A., van den Elst, H., Filippov, D.V., Blaser, H., Raz, E., Moens, C.B., et al. (2007). A role for Piwi and piRNAs in germ cell maintenance and transposon silencing in zebrafish. *Cell* 129, pp. 69–82.

Howell, C.Y., Bestor, T.H., Ding, F., Latham, K.E., Mertineit, C., Trasler, J.M., and Chaillet, J.R. (2001). Genomic imprinting disrupted by a maternal effect mutation in the Dnmt1 gene. *Cell* 104, pp. 829–38.

Hu, Y., Baud, V., Delhase, M., Zhang, P., Deerinck, T., Ellisman, M., Johnson, R., and Karin, M. (1999). Abnormal morphogenesis but intact IKK activation in mice lacking the IKKalpha subunit of IkappaB kinase. *Science* 284, pp. 316–20.

Hu, W., Feng, Z., Atwal, G.S., and Levine, A.J. (2008). p53: a new player in reproduction. *Cell Cycle* 7, pp. 848–52.

Hu, W., Feng, Z., Teresky, A.K., and Levine, A.J. (2007). p53 regulates maternal reproduction through LIF. *Nature* 450, pp. 721–24.

Huang, H., Rambaldi, I., Daniels, E., and Featherstone, M. (2003). Expression of the Wdr9 gene and protein products during mouse development. *Dev Dyn* 227, pp. 608–14.

Huang, P., and Schier, A.F. (2009). Dampened Hedgehog signaling but normal Wnt signaling in zebrafish without cilia. *Development* 136, pp. 3089–98.

Hussein, T.S., Thompson, J.G., and Gilchrist, R.B. (2006). Oocyte-secreted factors enhance oocyte developmental competence. *Dev Biol* 296, pp. 514–21.

Hutvagner, G., McLachlan, J., Pasquinelli, A.E., Balint, E., Tuschl, T., and Zamore, P.D. (2001). A cellular function for the RNA-interference enzyme Dicer in the maturation of the let-7 small temporal RNA. *Science* 293, pp. 834–38.

Hutvagner, G., and Zamore, P.D. (2002). A microRNA in a multiple-turnover RNAi enzyme complex. *Science* 297, pp. 2056–60.

Hutvagner, G., and Simard, M.J. (2008). Argonaute proteins: key players in RNA silencing. *Nat Rev Mol Cell Biol* 9, pp. 22–32.

Hyenne, V., Souilhol, C., Cohen-Tannoudji, M., Cereghini, S., Petit, C., Langa, F., Maro, B., and

Simmler, M.C. (2007). Conditional knock-out reveals that zygotic vezatin-null mouse embryos die at implantation. *Mech Dev* 124, pp. 449–62.

Ingledue, T.C., 3rd, Dominski, Z., Sanchez, R., Erkmann, J.A., and Marzluff, W.F. (2000). Dual role for the RNA-binding domain of *Xenopus laevis* SLBP1 in histone pre-mRNA processing. *RNA* 6, pp. 1635–48.

Ingraham, C.R., Kinoshita, A., Kondo, S., Yang, B., Sajan, S., Trout, K.J., Malik, M.I., Dunnwald, M., Goudy, S.L., Lovett, M., et al. (2006). Abnormal skin, limb and craniofacial morphogenesis in mice deficient for interferon regulatory factor 6 (Irf6). *Nat Genet* 38, pp. 1335–40.

Jansen, R.P. (2000). Germline passage of mitochondria: quantitative considerations and possible embryological sequelae. *Hum Reprod* 15 (Suppl 2), pp. 112–28.

Jedrusik, A., Bruce, A.W., Tan, M.H., Leong, D.E., Skamagki, M., Yao, M., and Zernicka-Goetz, M. (2010). Maternally and zygotically provided Cdx2 have novel and critical roles for early development of the mouse embryo. *Dev Biol* 344, pp. 66–78.

Jedrusik, A., Parfitt, D.E., Guo, G., Skamagki, M., Grabarek, J.B., Johnson, M.H., Robson, P., and Zernicka-Goetz, M. (2008). Role of Cdx2 and cell polarity in cell allocation and specification of trophectoderm and inner cell mass in the mouse embryo. *Genes Dev* 22, pp. 2692–706.

Jenny, A. (2010). Planar cell polarity signaling in the *Drosophila* eye. *Curr Top Dev Biol* 93, pp. 189–227.

Jenny, A., and Mlodzik, M. (2006). Planar cell polarity signaling: a common mechanism for cellular polarization. *Mt Sinai J Med* 73, pp. 738–50.

Jenuwein, T., and Allis, C.D. (2001). Translating the histone code. *Science* 293, pp. 1074–80.

Jessen, J.R., Topczewski, J., Bingham, S., Sepich, D.S., Marlow, F., Chandrasekhar, A., and Solnica-Krezel, L. (2002). Zebrafish trilobite identifies new roles for Strabismus in gastrulation and neuronal movements. *Nat Cell Biol* 4, pp. 610–15.

Jesuthasan, S. (1998). Furrow-associated microtubule arrays are required for the cohesion of zebrafish blastomeres following cytokinesis. *J Cell Sci* 111 (Pt 24), pp. 3695–703.

Jesuthasan, S., and Strahle, U. (1997). Dynamic microtubules and specification of the zebrafish embryonic axis. *Curr Biol* 7, pp. 31–42.

Johnson, M.H. (2009). From mouse egg to mouse embryo: polarities, axes, and tissues. Annu Rev Cell *Dev Biol* 25, pp. 483–512.

Johnson, M.H., Eager, D., Muggleton-Harris, A., and Grave, H.M. (1975). Mosaicism in organisation concanavalin A receptors on surface membrane of mouse egg. *Nature* 257, pp. 321–22.

Johnson, M.H., Maro, B., and Takeichi, M. (1986). The role of cell adhesion in the synchronization and orientation of polarization in 8-cell mouse blastomeres. *J Embryol Exp Morphol* 93, pp. 239–55.

Jordan, P., and Karess, R. (1997). Myosin light chain-activating phosphorylation sites are required for oogenesis in *Drosophila*. *J Cell Biol* 139, pp. 1805–19.

Joshi, S., Davies, H., Sims, L.P., Levy, S.E., and Dean, J. (2007). Ovarian gene expression in the absence of FIGLA, an oocyte-specific transcription factor. BMC *Dev Biol* 7, p. 67.

Kai, T., Williams, D., and Spradling, A.C. (2005). The expression profile of purified *Drosophila* germline stem cells. *Dev Biol* 283, pp. 486–502.

Kane, D., and Adams, R. (2002). Life at the edge: epiboly and involution in the zebrafish. *Results Probl Cell Differ* 40, pp. 117–35.

Kane, D.A., and Kimmel, C.B. (1993). The zebrafish midblastula transition. *Development* 119, pp. 447–56.

Kane, D.A., McFarland, K.N., and Warga, R.M. (2005). Mutations in half baked/E-cadherin block cell behaviors that are necessary for teleost epiboly. *Development* 132, pp. 1105–16.

Kaneda, M., Tang, F., O'Carroll, D., Lao, K., and Surani, M.A. (2009). Essential role for Argonaute2 protein in mouse oogenesis. *Epigenetics Chromatin* 2, p. 9.

Kang, H.J., Feng, Z., Sun, Y., Atwal, G., Murphy, M.E., Rebbeck, T.R., Rosenwaks, Z., Levine, A.J., and Hu, W. (2009). Single-nucleotide polymorphisms in the p53 pathway regulate fertility in humans. *Proc Natl Acad Sci U S A* 106, pp. 9761–66.

Kanzler, B., Haas-Assenbaum, A., Haas, I., Morawiec, L., Huber, E., and Boehm, T. (2003). Morpholino oligonucleotide-triggered knockdown reveals a role for maternal E-cadherin during early mouse development. *Mech Dev* 120, pp. 1423–32.

Kato, Y., Kaneda, M., Hata, K., Kumaki, K., Hisano, M., Kohara, Y., Okano, M., Li, E., Nozaki, M., and Sasaki, H. (2007). Role of the Dnmt3 family in de novo methylation of imprinted and repetitive sequences during male germ cell development in the mouse. *Hum Mol Genet* 16, 2272–80.

Kedde, M., Strasser, M.J., Boldajipour, B., Oude Vrielink, J.A., Slanchev, K., le Sage, C., Nagel, R., Voorhoeve, P.M., van Duijse, J., Orom, U.A., et al. (2007). RNA-binding protein Dnd1 inhibits microRNA access to target mRNA. *Cell* 131, pp. 1273–86.

Keegan, B.R., Feldman, J.L., Begemann, G., Ingham, P.W., and Yelon, D. (2005). Retinoic acid signaling restricts the cardiac progenitor pool. *Science* 307, pp. 247–9.

Kehler, J., Tolkunova, E., Koschorz, B., Pesce, M., Gentile, L., Boiani, M., Lomeli, H., Nagy, A., McLaughlin, K.J., Scholer, H.R., et al. (2004). Oct4 is required for primordial germ cell survival. *EMBO Rep* 5, pp. 1078–83.

Keller, R., and Hardin, J. (1987). Cell behaviour during active cell rearrangement: evidence and speculations. *J Cell Sci* Suppl 8, pp. 369–93.

Kelly, G.M., Greenstein, P., Erezyilmaz, D.F., and Moon, R.T. (1995). Zebrafish wnt8 and wnt8b share a common activity but are involved in distinct developmental pathways. *Development* 121, pp. 1787–99.

Kelly, C., Chin, A.J., and Weinberg, E.S. (1998). ichabod, a zebrafish maternal effect mutation affecting establishment of the organizer. Paper presented at: Zebrafish Development and Genetics (Cold Spring Harbor Laboratory, Cold Spring Harbor, NY).

Kemphues, K., and Strome, S. (1997). Fertilization and establishment of polarity in the embryo. (Woodbury, Cold Spring Harbor Laboratory Press).

Kennerdell, J.R., Yamaguchi, S., and Carthew, R.W. (2002). RNAi is activated during *Drosophila* oocyte maturation in a manner dependent on aubergine and spindle-E. *Genes Dev* 16, pp. 1884–9.

Ketting, R.F., Fischer, S.E., Bernstein, E., Sijen, T., Hannon, G.J., and Plasterk, R.H. (2001). Dicer functions in RNA interference and in synthesis of small RNA involved in developmental timing in *C. elegans. Genes Dev* 15, pp. 2654–9.

Kijima, M., Yoshida, M., Sugita, K., Horinouchi, S., and Beppu, T. (1993). Trapoxin, an antitumor cyclic tetrapeptide, is an irreversible inhibitor of mammalian histone deacetylase. *J Biol Chem* 268, pp. 22429–35.

Kilian, B., Mansukoski, H., Barbosa, F.C., Ulrich, F., Tada, M., and Heisenberg, C.P. (2003). The role of Ppt / Wnt5 in regulating cell shape and movement during zebrafish gastrulation. *Mech Dev* 120, pp. 467–76.

Kim-Ha, J., Kim, J., and Kim, Y.J. (1999). Requirement of RBP9, a *Drosophila* Hu homolog, for regulation of cystocyte differentiation and oocyte determination during oogenesis. *Mol Cell Biol* 19, pp. 2505–14.

Kim, C.H., Oda, T., Itoh, M., Jiang, D., Artinger, K.B., Chandrasekharappa, S.C., Driever, W., and Chitnis, A.B. (2000). Repressor activity of Headless/Tcf3 is essential for vertebrate head formation. *Nature* 407, pp. 913–16.

Kim, D.Y., and Roy, R. (2006). Cell cycle regulators control centrosome elimination during oogenesis in *Caenorhabditis elegans. J Cell Biol* 174, pp. 751–57.

Kimmel, C.B., Warga, R.M., and Schilling, T.F. (1990). Origin and organization of the zebrafish fate map. *Development* 108, pp. 581–94.

King, M.L., Messitt, T.J., and Mowry, K.L. (2005). Putting RNAs in the right place at the right time: RNA localization in the frog oocyte. *Biol Cell* 97, pp. 19–33.

Kishimoto, Y., Lee, K.H., Zon, L., Hammerschmidt, M., and Schulte-Merker, S. (1997). The molecular nature of zebrafish swirl: BMP2 function is essential during early dorsoventral patterning. *Development* 124, pp. 4457–66.

Kishimoto, Y., Koshida, S., Furutani-Seiki, M., and Kondoh, H. (2004). Zebrafish maternal-effect mutations causing cytokinesis defect without affecting mitosis or equatorial vasa deposition. *Mech Dev* 121, pp. 79–89.

Klein, P.S., and Melton, D.A. (1996). A molecular mechanism for the effect of lithium on development. *Proc Natl Acad Sci USA* 93, pp. 8455–9.

Kloc, M., and Etkin, L.D. (1994). Delocalization of Vg1 mRNA from the vegetal cortex in *Xenopus* oocytes after destruction of Xlsirt RNA. *Science* 265, pp. 1101–3.

Kloc, M., Bilinski, S., Chan, A.P., Allen, L.H., Zearfoss, N.R., and Etkin, L.D. (2001). RNA localization and germ cell determination in *Xenopus. Int Rev Cytol* 203, pp. 63–91.

Kloc, M., Dougherty, M.T., Bilinski, S., Chan, A.P., Brey, E., King, M.L., Patrick, C.W., Jr., and Etkin, L.D. (2002). Three-dimensional ultrastructural analysis of RNA distribution within germinal granules of *Xenopus. Dev Biol* 241, pp. 79–93.

Kloc, M., Bilinski, S., Dougherty, M.T., Brey, E.M., and Etkin, L.D. (2004). Formation, architecture and polarity of female germline cyst in *Xenopus. Dev Biol* 266, pp. 43–61.

Kloc, M., Bilinski, S., and Etkin, L.D. (2004a). The Balbiani body and germ cell determinants: 150 years later. *Curr Top Dev Biol* 59, pp. 1–36.

Kloc, M., and Etkin, L.D. (2005). RNA localization mechanisms in oocytes. *J Cell Sci* 118, pp. 269–82.

Kloc, M., Jaglarz, M., Dougherty, M., Stewart, M.D., Nel-Themaat, L., and Bilinski, S. (2008). Mouse early oocytes are transiently polar: three-dimensional and ultrastructural analysis. *Exp Cell Res* 314, pp. 3245–54.

Knaut, H., Pelegri, F., Bohmann, K., Schwarz, H., and Nusslein-Volhard, C. (2000). Zebrafish vasa RNA but not its protein is a component of the germ plasm and segregates asymmetrically before germline specification. *J Cell Biol* 149, pp. 875–88.

Knight, S.W., and Bass, B.L. (2001). A role for the RNase III enzyme DCR-1 in RNA interference and germ line development in *Caenorhabditis elegans. Science* 293, pp. 2269–71.

Kodama, Y., Rothman, J.H., Sugimoto, A., and Yamamoto, M. (2002). The stem-loop binding protein CDL-1 is required for chromosome condensation, progression of cell death and morphogenesis in *Caenorhabditis elegans. Development* 129, pp. 187–96.

Kofron, M., Birsoy, B., Houston, D., Tao, Q., Wylie, C., and Heasman, J. (2007). Wnt11/beta-catenin signaling in both oocytes and early embryos acts through LRP6-mediated regulation of axin. *Development* 134, pp. 503–13.

Kofron, M., Klein, P., Zhang, F., Houston, D.W., Schaible, K., Wylie, C., and Heasman, J. (2001). The role of maternal axin in patterning the *Xenopus* embryo. *Dev Biol* 237, pp. 183–201.

Kofron, M., Puck, H., Standley, H., Wylie, C., Old, R., Whitman, M., and Heasman, J. (2004). New roles for FoxH1 in patterning the early embryo. *Development* 131, pp. 5065–78.

Kofron, M., Spagnuolo, A., Klymkowsky, M., Wylie, C., and Heasman, J. (1997). The roles of maternal alpha-catenin and plakoglobin in the early *Xenopus* embryo. *Development* 124, pp. 1553–60.

Kohn, A.D., and Moon, R.T. (2005). Wnt and calcium signaling: beta-catenin independent pathways. *Cell Calcium* 38, pp. 439–46.

Koken, M.H., Smit, E.M., Jaspers-Dekker, I., Oostra, B.A., Hagemeijer, A., Bootsma, D., and Hoeijmakers, J.H. (1992). Localization of two human homologs, HHR6A and HHR6B, of the yeast DNA repair gene RAD6 to chromosomes Xq24-q25 and 5q23-q31. *Genomics* 12, pp. 447–53.

Kosaka, K., Kawakami, K., Sakamoto, H., and Inoue, K. (2007). Spatiotemporal localization of germ plasm RNAs during zebrafish oogenesis. *Mech Dev* 124, pp. 279–89.

Kotani, T., and Kawakami, K. (2008). Misty somites, a maternal effect gene identified by transposon-mediated insertional mutagenesis in zebrafish that is essential for the somite boundary maintenance. *Dev Biol* 316, pp. 383–96.

Kramer, C., Mayr, T., Nowak, M., Schumacher, J., Runke, G., Bauer, H., Wagner, D.S., Schmid, B., Imai, Y., Talbot, W.S., et al. (2002). Maternally supplied Smad5 is required for ventral specification in zebrafish embryos prior to zygotic Bmp signaling. *Dev Biol* 250, pp. 263–79.

Krauss, S., Korzh, V., Fjose, A., and Johansen, T. (1992). Expression of four zebrafish wnt-related genes during embryogenesis. *Development* 116, pp. 249–59.

Krogan, N.J., Kim, M., Tong, A., Golshani, A., Cagney, G., Canadien, V., Richards, D.P., Beattie, B.K., Emili, A., Boone, C., et al. (2003). Methylation of histone H3 by Set2 in *Saccharomyces cerevisiae* is linked to transcriptional elongation by RNA polymerase II. *Mol Cell Biol* 23, pp. 4207–18.

Kruh, J. (1982). Effects of sodium butyrate, a new pharmacological agent, on cells in culture. *Mol Cell Biochem* 42, pp. 65–82.

Ku, M., and Melton, D.A. (1993). Xwnt-11: a maternally expressed *Xenopus* wnt gene. *Development* 119, pp. 1161–73.

Kumar, T.R., Wang, Y., Lu, N., and Matzuk, M.M. (1997). Follicle stimulating hormone is required for ovarian follicle maturation but not male fertility. *Nat Genet* 15, pp. 201–04.

Kunwar, P.S., Zimmerman, S., Bennett, J.T., Chen, Y., Whitman, M., and Schier, A.F. (2003). Mixer/Bon and FoxH1/Sur have overlapping and divergent roles in Nodal signaling and mesendoderm induction. *Development* 130, pp. 5589–99.

Kuramochi-Miyagawa, S., Kimura, T., Yomogida, K., Kuroiwa, A., Tadokoro, Y., Fujita, Y., Sato, M., Matsuda, Y., and Nakano, T. (2001). Two mouse piwi-related genes: miwi and mili. *Mech Dev* 108, pp. 121–33.

Kuramochi-Miyagawa, S., Kimura, T., Ijiri, T.W., Isobe, T., Asada, N., Fujita, Y., Ikawa, M., Iwai, N., Okabe, M., Deng, W., et al. (2004). Mili, a mammalian member of piwi family gene, is essential for spermatogenesis. *Development* 131, pp. 839–49.

Kuramochi-Miyagawa, S., Watanabe, T., Gotoh, K., Totoki, Y., Toyoda, A., Ikawa, M., Asada, N., Kojima, K., Yamaguchi, Y., Ijiri, T.W., et al. (2008). DNA methylation of retrotransposon genes is regulated by Piwi family members MILI and MIWI2 in murine fetal testes. *Genes Dev* 22, pp. 908–17.

Kurihara, Y., Kawamura, Y., Uchijima, Y., Amamo, T., Kobayashi, H., Asano, T., and Kurihara, H. (2008). Maintenance of genomic methylation patterns during preimplantation development requires the somatic form of DNA methyltransferase 1. *Dev Biol* 313, pp. 335–46.

Labbe, J.C., Capony, J.P., Caput, D., Cavadore, J.C., Derancourt, J., Kaghad, M., Lelias, J.M., Picard, A., and Doree, M. (1989). MPF from starfish oocytes at first meiotic metaphase is a heterodimer containing one molecule of cdc2 and one molecule of cyclin B. *EMBO J* 8, pp. 3053–8.

Lai, W.S., Carballo, E., Strum, J.R., Kennington, E.A., Phillips, R.S., and Blackshear, P.J. (1999). Evidence that tristetraprolin binds to AU-rich elements and promotes the deadenylation and destabilization of tumor necrosis factor alpha mRNA. *Mol Cell Biol* 19, pp. 4311–23.

Lai, W.S., Carballo, E., Thorn, J.M., Kennington, E.A., and Blackshear, P.J. (2000). Interactions of CCCH zinc finger proteins with mRNA. Binding of tristetraprolin-related zinc finger proteins to Au-rich elements and destabilization of mRNA. *J Biol Chem* 275, pp. 17827–37.

Lanzotti, D.J., Kaygun, H., Yang, X., Duronio, R.J., and Marzluff, W.F. (2002). Developmental control of histone mRNA and dSLBP synthesis during *Drosophila* embryogenesis and the role of dSLBP in histone mRNA 3' end processing in vivo. *Mol Cell Biol* 22, pp. 2267–82.

Larabell, C.A., Torres, M., Rowning, B.A., Yost, C., Miller, J.R., Wu, M., Kimelman, D., and Moon, R.T. (1997). Establishment of the dorso-ventral axis in *Xenopus* embryos is presaged by early asymmetries in beta-catenin that are modulated by the Wnt signaling pathway. *J Cell Biol* 136, pp. 1123–36.

Larue, L., Ohsugi, M., Hirchenhain, J., and Kemler, R. (1994). E-cadherin null mutant embryos fail to form a trophectoderm epithelium. *Proc Natl Acad Sci U S A* 91, pp. 8263–67.

Lasko, P.F. (1992). Molecular movements in oocyte patterning and pole cell differentiation. *Bioessays* 14, pp. 507–12.

Lau, N.C., Ohsumi, T., Borowsky, M., Kingston, R.E., and Blower, M.D. (2009a). Systematic and single cell analysis of *Xenopus* Piwi-interacting RNAs and Xiwi. *EMBO J* 28, pp. 2945–58.

Lau, N.C., Robine, N., Martin, R., Chung, W.J., Niki, Y., Berezikov, E., and Lai, E.C. (2009b). Abundant primary piRNAs, endo-siRNAs, and microRNAs in a *Drosophila* ovary cell line. *Genome Res* 19, pp. 1776–85.

Lau, N.C., Seto, A.G., Kim, J., Kuramochi-Miyagawa, S., Nakano, T., Bartel, D.P., and Kingston, R.E. (2006). Characterization of the piRNA complex from rat testes. *Science* 313, pp. 363–7.

Lavoie, C.A., Ohlstein, B., and McKearin, D.M. (1999). Localization and function of Bam protein require the benign gonial cell neoplasm gene product. *Dev Biol* 212, pp. 405–13.

Leader, B., Lim, H., Carabatsos, M.J., Harrington, A., Ecsedy, J., Pellman, D., Maas, R., and Leder, P. (2002). Formin-2, polyploidy, hypofertility and positioning of the meiotic spindle in mouse oocytes. *Nat Cell Biol* 4, pp. 921–28.

Lee, J., Miyano, T., and Moor, R.M. (2000). Spindle formation and dynamics of gamma-tubulin and nuclear mitotic apparatus protein distribution during meiosis in pig and mouse oocytes. *Biol Reprod* 62, pp. 1184–92.

Lehmann, R., and Nusslein-Volhard, C. (1986). Abdominal segmentation, pole cell formation, and embryonic polarity require the localized activity of oskar, a maternal gene in *Drosophila. Cell* 47, pp. 141–52.

Lekven, A.C., Thorpe, C.J., Waxman, J.S., and Moon, R.T. (2001). Zebrafish wnt8 encodes two wnt8 proteins on a bicistronic transcript and is required for mesoderm and neurectoderm patterning. *Dev Cell* 1, pp. 103–14.

Leon, A., and McKearin, D. (1999). Identification of TER94, an AAA ATPase protein, as a Bam-dependent component of the *Drosophila* fusome. *Mol Biol Cell* 10, pp. 3825–34.

Lessard, C., Pendola, J.K., Hartford, S.A., Schimenti, J.C., Handel, M.A., and Eppig, J.J. (2004). New mouse genetic models for human contraceptive development. *Cytogenet Genome Res* 105, pp. 222–27.

Leung, C.F., Webb, S.E., and Miller, A.L. (1998). Calcium transients accompany ooplasmic segregation in zebrafish embryos. *Dev Growth Differ* 40, pp. 313–26.

Leung, C.F., Webb, S.E., and Miller, A.L. (2000). On the mechanism of ooplasmic segregation in single-cell zebrafish embryos. *Dev Growth Differ* 42, pp. 29–40.

Leung, T., Soll, I., Arnold, S.J., Kemler, R., and Driever, W. (2003). Direct binding of Lef1 to sites in the boz promoter may mediate pre-midblastula-transition activation of boz expression. *Dev Dyn* 228, pp. 424–32.

Lewis, A., Mitsuya, K., Umlauf, D., Smith, P., Dean, W., Walter, J., Higgins, M., Feil, R., and Reik, W. (2004). Imprinting on distal chromosome 7 in the placenta involves repressive histone methylation independent of DNA methylation. *Nat Genet* 36, pp. 1291–5.

Li, M.G., Serr, M., Newman, E.A., and Hays, T.S. (2004). The *Drosophila* tctex-1 light chain is dispensable for essential cytoplasmic dynein functions but is required during spermatid differentiation. *Mol Biol Cell* 15, pp. 3005–14.

Lewis, R.A., Gagnon, J.A., and Mowry, K.L. (2008). PTB/hnRNP I is required for RNP remodeling during RNA localization in *Xenopus* oocytes. *Mol Cell Biol* 28, *pp.* 678–86.

Li, M.G., Serr, M., Edwards, K., Ludmann, S., Yamamoto, D., Tilney, L.G., Field, C.M., and Hays, T.S. (1999). Filamin is required for ring canal assembly and actin organization during Drosophila oogenesis. *J Cell Biol* 146, pp. 1061–74.

Li, X., and Leder, P. (2007). Identifying genes preferentially expressed in undifferentiated embryonic stem cells. *BMC Cell Biol* 8, p. 37.

Li, L., Baibakov, B., and Dean, J. (2008a). A subcortical maternal complex essential for preimplantation mouse embryogenesis. *Dev Cell* 15, pp. 416–25.

Li, X., Ito, M., Zhou, F., Youngson, N., Zuo, X., Leder, P., and Ferguson-Smith, A.C. (2008b). A maternal-zygotic effect gene, Zfp57, maintains both maternal and paternal imprints. *Dev Cell* 15, pp. 547–57.

Liang, L., Soyal, S.M., and Dean, J. (1997). FIGalpha, a germ cell specific transcription factor involved in the coordinate expression of the zona pellucida genes. *Development* 124, pp. 4939–47.

Lighthouse, D.V., Buszczak, M., and Spradling, A.C. (2008). New components of the *Drosophila* fusome suggest it plays novel roles in signaling and transport. *Dev Biol* 317, pp. 59–71.

Lilly, M.A., de Cuevas, M., and Spradling, A.C. (2000). Cyclin A associates with the fusome during germline cyst formation in the *Drosophila* ovary. *Dev Biol* 218, pp. 53–63.

Lin, H., Yue, L., and Spradling, A.C. (1994). The Drosophila fusome, a germline- specific organelle, contains membrane skeletal proteins and functions in cyst formation. *Development* 120, pp. 947–56.

Lin, H., and Spradling, A.C. (1995). Fusome asymmetry and oocyte determination in Drosophila. *Dev Genet* 16, pp. 6–12.

Lindeman, R.E., and Pelegri, F. (2009). Vertebrate maternal-effect genes: insights into fertilization, early cleavage divisions, and germ cell determinant localization from studies in the zebrafish. *Mol Reprod Dev.*

Liu, Z., Xie, T., and Steward, R. (1999). Lis1, the Drosophila homolog of a human lissencephaly disease gene, is required for germline cell division and oocyte differentation. *Development* 126, pp. 4477–88.

Lloyd, B., Tao, Q., Lang, S., and Wylie, C. (2005). Lysophosphatidic acid signaling controls cortical actin assembly and cytoarchitecture in *Xenopus* embryos. *Development* 132, pp. 805–16.

Lucifero, D., Mann, M.R., Bartolomei, M.S., and Trasler, J.M. (2004). Genespecific timing and epigenetic memory in oocyte imprinting. *Hum Mol Genet* 13, pp. 839–49.

Lund, E., Liu, M., Hartley, R.S., Sheets, M.D., and Dahlberg, J.E. (2009). Deadenylation of maternal mRNAs mediated by miR-427 in *Xenopus laevis* embryos. *RNA* 15, pp. 2351–63.

Lunde, K., Belting, H.G., and Driever, W. (2004). Zebrafish pou5f1/pou2, homolog of mammalian Oct4, functions in the endoderm specification cascade. *Curr Biol* 14, pp. 48–55.

Luschnig, S., Moussian, B., Krauss, J., Desjeux, I., Perkovic, J., and Nusslein-Volhard, C. (2004). An F1 genetic screen for maternal-effect mutations affecting embryonic pattern formation in *Drosophila melanogaster. Genetics* 167, pp. 325–42.

Lutz, B., Schmid, W., Niehrs, C., and Schutz, G. (1999). Essential role of CREB family proteins during *Xenopus* embryogenesis. *Mech Dev* 88, pp. 55–66.

Lykke-Andersen, K., Gilchrist, M.J., Grabarek, J.B., Das, P., Miska, E., and Zernicka-Goetz, M. (2008). Maternal Argonaute 2 is essential for early mouse development at the maternal-zygotic transition. *Mol Biol Cell* 19, pp. 4383–92.

Lyman Gingerich, J., Westfall, T.A., Slusarski, D.C., and Pelegri, F. (2005). Hecate, a zebrafish maternal effect gene, affects dorsal organizer induction and intracellular calcium transient frequency. *Dev Biol* 286, pp. 427–39.

Ma, J., Svoboda, P., Schultz, R.M., and Stein, P. (2001). Regulation of zygotic gene activation in the preimplantation mouse embryo: global activation and repression of gene expression. *Biol Reprod* 64, pp. 1713–21.

Ma, C., Benink, H.A., Cheng, D., Montplaisir, V., Wang, L., Xi, Y., Zheng, P.P., Bement, W.M., and Liu, X.J. (2006a). Cdc42 activation couples spindle positioning to first polar body formation in oocyte maturation. *Curr Biol* 16, pp. 214–20.

Ma, J., Zeng, F., Schultz, R.M., and Tseng, H. (2006b). Basonuclin: a novel mammalian maternal-effect gene. *Development* 133, pp. 2053–62.

Mackay, D.J., Callaway, J.L., Marks, S.M., White, H.E., Acerini, C.L., Boonen, S.E., Dayanikli, P., Firth, H.V., Goodship, J.A., Haemers, A.P., et al. (2008). Hypomethylation of multiple imprinted loci in individuals with transient neonatal diabetes is associated with mutations in ZFP57. *Nat Genet* 40, pp. 949–51.

Maegawa, S., Yasuda, K., and Inoue, K. (1999). Maternal mRNA localization of zebrafish DAZ-like gene. *Mech Dev* 81, pp. 223–26.

Mager, J., Schultz, R.M., Brunk, B.P., and Bartolomei, M.S. (2006). Identification of candidate maternal-effect genes through comparison of multiple microarray data sets. *Mamm Genome* 17, pp. 941–9.

Majumder, S., and DePamphilis, M.L. (1994). Requirements for DNA transcription and replication at the beginning of mouse development. *J Cell Biochem* 55, pp. 59–68.

Makita, R., Mizuno, T., Kuroiwa, A., Koshida, S., and Takeda, H. (1998). Zebrafish wnt11: pattern and regulation of the expression by the yolk cell and no tail activity. *Mech Dev* 71, pp. 165–76.

Malynn, B.A., de Alboran, I.M., O'Hagan, R.C., Bronson, R., Davidson, L., DePinho, R.A., and Alt, F.W. (2000). N-myc can functionally replace c-myc in murine development, cellular growth, and differentiation. *Genes Dev* 14, pp. 1390–99.

Marikawa, Y., Li, Y., and Elinson, R.P. (1997). Dorsal determinants in the *Xenopus* egg are firmly associated with the vegetal cortex and behave like activators of the Wnt pathway. *Dev Biol* 191, pp. 69–79.

Marinos, E., and Billett, F.S. (1981). Mitochondrial number, cytochrome oxidase and succinic dehydrogenase activity in *Xenopus laevis* oocytes. *J Embryol Exp Morphol* 62, pp. 395–409.

Marlow, F.L., and Mullins, M.C. (2008). Bucky ball functions in Balbiani body assembly and animal–vegetal polarity in the oocyte and follicle cell layer in zebrafish. *Dev Biol* 321, pp. 40–50.

Marnef, A., Sommerville, J., and Ladomery, M.R. (2009). RAP55: insights into an evolutionarily conserved protein family. *Int J Biochem Cell Biol* 41, pp. 977–81.

Masui, Y., and Markert, C.L. (1971). Cytoplasmic control of nuclear behavior during meiotic maturation of frog oocytes. *J Exp Zool* 177, pp. 129–45.

Masui, Y. (2001). From oocyte maturation to the in vitro cell cycle: the history of discoveries of Maturation-Promoting Factor (MPF) and Cytostatic Factor (CSF). *Differentiation* 69, pp. 1–17.

Mathavan, S., Lee, S.G., Mak, A., Miller, L.D., Murthy, K.R., Govindarajan, K.R., Tong, Y., Wu, Y.L., Lam, S.H., Yang, H., et al. (2005). Transcriptome analysis of zebrafish embryogenesis using microarrays. *PLoS Genet* 1, pp. 260–76.

Mathe, E., Inoue, Y.H., Palframan, W., Brown, G., and Glover, D.M. (2003). Orbit/Mast, the CLASP orthologue of *Drosophila*, is required for asymmetric stem cell and cystocyte divisions and development of the polarised microtubule network that interconnects oocyte and nurse cells during oogenesis. *Development* 130, pp. 901–5.

Matuszewski, B., Ciechomski, K., and Kloc, M. (1999). Extrachromosomal rDNA and polarity of pro-oocytes during ovary development in *Creophilus maxillosus* (Coleoptera, Staphylinidae). *Folia Histochem Cytobiol* 37, pp. 179–90.

Matzuk, M.M., Burns, K.H., Viveiros, M.M., and Eppig, J.J. (2002). Intercellular communication in the mammalian ovary: oocytes carry the conversation. *Science* 296, pp. 2178–80.

McGaughey, R.W., and Capco, D.G. (1989). Specialized cytoskeletal elements in mammalian eggs: structural and biochemical evidence for their composition. *Cell Motil Cytoskeleton* 13, pp. 104–11.

McKearin, D., and Ohlstein, B. (1995). A role for the *Drosophila* bag-of-marbles protein in the differentiation of cystoblasts from germline stem cells. *Development* 121, pp. 2937–47.

McLaren, A. (2003). Primordial germ cells in the mouse. *Dev Biol* 262, pp. 1–15.

McMahon, A.P., and Moon, R.T. (1989). Ectopic expression of the protooncogene int-1 in *Xenopus* embryos leads to duplication of the embryonic axis. *Cell* 58, pp. 1075–84.

McNeilly, J.R., Saunders, P.T., Taggart, M., Cranfield, M., Cooke, H.J., and McNeilly, A.S. (2000). Loss of oocytes in Dazl knockout mice results in maintained ovarian steroidogenic function but altered gonadotropin secretion in adult animals. *Endocrinology* 141, pp. 4284–94.

Megosh, H.B., Cox, D.N., Campbell, C., and Lin, H. (2006). The role of PIWI and the miRNA machinery in *Drosophila* germline determination. *Curr Biol* 16, pp. 1884–94.

Mei, W., Lee, K.W., Marlow, F.L., Miller, A.L., and Mullins, M.C. (2009). hnRNP I is required to generate the Ca2+ signal that causes egg activation in zebrafish. *Development* 136, pp. 3007–17.

Melton, D.A. (1987). Translocation of a localized maternal mRNA to the vegetal pole of *Xenopus* oocytes. *Nature* 328, pp. 80–82.

Melton, D.A. (1991). Pattern formation during animal development. *Science* 252, pp. 234–41.

Melton, D.A., Ruiz i Altaba, A., Yisraeli, J., and Sokol, S. (1989). Localization of mRNA and axis formation during Xenopus embryogenesis. *Ciba Found Symp* 144, pp. 16–29; discussion 29–36, pp. 92–18.

Memili, E., and First, N.L. (2000). Zygotic and embryonic gene expression in cow: a review of timing and mechanisms of early gene expression as compared with other species. *Zygote* 8, pp. 87–96.

Meno, C., Gritsman, K., Ohishi, S., Ohfuji, Y., Heckscher, E., Mochida, K., Shimono, A., Kondoh, H., Talbot, W.S., Robertson, E.J., et al. (1999). Mouse Lefty2 and zebrafish antivin are feedback inhibitors of nodal signaling during vertebrate gastrulation. *Mol Cell* 4, pp. 287–98.

Meric, F., Matsumoto, K., and Wolffe, A.P. (1997). Regulated unmasking of in vivo synthesized maternal mRNA at oocyte maturation. A role for the chaperone nucleoplasmin. *J Biol Chem* 272, pp. 12840–46.

Messitt, T.J., Gagnon, J.A., Kreiling, J.A., Pratt, C.A., Yoon, Y.J., and Mowry, K.L. (2008). Multiple kinesin motors coordinate cytoplasmic RNA transport on a subpopulation of microtubules in *Xenopus* oocytes. *Dev Cell* 15, pp. 426–36.

Metchat, A., Akerfelt, M., Bierkamp, C., Delsinne, V., Sistonen, L., Alexandre, H., and Christians, E.S. (2009). Mammalian heat shock factor 1 is essential for oocyte meiosis and directly regulates Hsp90alpha expression. *J Biol Chem* 284, pp. 9521–8.

Mikkelsen, T.S., Ku, M., Jaffe, D.B., Issac, B., Lieberman, E., Giannoukos, G., Alvarez, P., Brockman, W., Kim, T.K., Koche, R.P., et al. (2007). Genome-wide maps of chromatin state in pluripotent and lineage-committed cells. *Nature* 448, pp. 553–60.

Miller, J.R., Hocking, A.M., Brown, J.D., and Moon, R.T. (1999). Mechanism and function of signal transduction by the Wnt/beta-catenin and Wnt/Ca2+ pathways. *Oncogene* 18, pp. 7860–72.

Minakhina, S., and Steward, R. (2005). Axes formation and RNA localization. *Curr Opin Genet Dev* 15, pp. 416–21.

Mir, A., and Heasman, J. (2008). How the mother can help: studying maternal Wnt signaling by anti-sense-mediated depletion of maternal mRNAs and the host transfer technique. *Methods Mol Biol* 469, pp. 417–29.

Mishima, Y., Giraldez, A.J., Takeda, Y., Fujiwara, T., Sakamoto, H., Schier, A.F., and Inoue, K. (2006). Differential regulation of germline mRNAs in soma and germ cells by zebrafish miR-430. *Curr Biol* 16, pp. 2135–42.

Misirlioglu, M., Page, G.P., Sagirkaya, H., Kaya, A., Parrish, J.J., First, N.L., and Memili, E. (2006). Dynamics of global transcriptome in bovine matured oocytes and preimplantation embryos. *Proc Natl Acad Sci U S A* 103, pp. 18905–10.

Mita, K., and Yamashita, M. (2000). Expression of *Xenopus* Daz-like protein during gametogenesis and embryogenesis. *Mech Dev* 94, pp. 251–5.

Miyara, F., Migne, C., Dumont-Hassan, M., Le Meur, A., Cohen-Bacrie, P., Aubriot, F.X., Glissant, A., Nathan, C., Douard, S., Stanovici, A., et al. (2003). Chromatin configuration and transcriptional control in human and mouse oocytes. *Mol Reprod Dev* 64, pp. 458–70.

Mizuno, T., Yamaha, E., Kuroiwa, A., and Takeda, H. (1999). Removal of vegetal yolk causes dorsal deficiencies and impairs dorsal-inducing ability of the yolk cell in zebrafish. *Mech Dev* 81, pp. 35–47.

Mizuno, T., Yamaha, E., Wakahara, M., Kuroiwa, A., and Takeda, H. (1996). Mesoderm induction in zebrafish. *Nature* 383, pp. 131–32.

Mlodzik, M. (2002). Planar cell polarization: do the same mechanisms regulate *Drosophila* tissue polarity and vertebrate gastrulation? *Trends Genet* 18, pp. 564–71.

Moon, R.T., Campbell, R.M., Christian, J.L., McGrew, L.L., Shih, J. and Fraser, S. (1993). Xwnt -5A: a maternal Wnt that affects morphogenetic movements after overexpression in embryos of *Xenopus laevis*. *Development* 119, pp. 91–111.

Moore, G.D., Ayabe, T., Visconti, P.E., Schultz, R.M., and Kopf, G.S. (1994). Roles of heterotrimeric and monomeric G proteins in sperm-induced activation of mouse eggs. *Development* 120, pp. 3313–23.

Morham, S.G., Kluckman, K.D., Voulomanos, N., and Smithies, O. (1996). Targeted disruption of the mouse topoisomerase I gene by camptothecin selection. *Mol Cell Biol* 16, pp. 6804–9.

Morris, J.Z., Hong, A., Lilly, M.A., and Lehmann, R. (2005). twin, a CCR4 homolog, regulates cyclin poly(A) tail length to permit *Drosophila* oogenesis. *Development* 132, pp. 1165–74.

Motosugi, N., Dietrich, J.E., Polanski, Z., Solter, D., and Hiiragi, T. (2006). Space asymmetry directs preferential sperm entry in the absence of polarity in the mouse oocyte. *PLoS Biol* 4, p. e135.

Mukai, M., Kitadate, Y., Arita, K., Shigenobu, S., and Kobayashi, S. (2006). Expression of meiotic genes in the germline progenitors of *Drosophila* embryos. *Gene Expr Patterns* 6, pp. 256–66.

Muller, H.A. (2001). Of mice, frogs and flies: generation of membrane asymmetries in early development. *Dev Growth Differ* 43, pp. 327–42.

Mullins, M.C., Hammerschmidt, M., Haffter, P., and Nusslein-Volhard, C. (1994). Large-scale mutagenesis in the zebrafish: in search of genes controlling development in a vertebrate. *Current Biology* 4, pp. 189–202.

Murakami, M.S., Copeland, T.D., and Vande Woude, G.F. (1999). Mos positively regulates Xe-Wee1 to lengthen the first mitotic cell cycle of *Xenopus*. *Genes Dev* 13, pp. 620–31.

Murchison, E.P., Stein, P., Xuan, Z., Pan, H., Zhang, M.Q., Schultz, R.M., and Hannon, G.J. (2007). Critical roles for Dicer in the female germline. *Genes Dev* 21, pp. 682–93.

Na, J., and Zernicka-Goetz, M. (2006). Asymmetric positioning and organization of the meiotic spindle of mouse oocytes requires CDC42 function. *Curr Biol* 16, pp. 1249–54.

Nakamura, T., Arai, Y., Umehara, H., Masuhara, M., Kimura, T., Taniguchi, H., Sekimoto, T., Ikawa, M., Yoneda, Y., Okabe, M., et al. (2007). PGC7/Stella protects against DNA demethylation in early embryogenesis. *Nat Cell Biol* 9, pp. 64–71.

Nance, J. (2005). PAR proteins and the establishment of cell polarity during *C. elegans* development. *Bioessays* 27, pp. 126–35.

Narbonne-Reveau, K., Besse, F., Lamour-Isnard, C., Busson, D., and Pret, A.M. (2006). fused regulates germline cyst mitosis and differentiation during *Drosophila* oogenesis. *Mech Dev* 123, pp. 197–209.

Narducci, M.G., Fiorenza, M.T., Kang, S.M., Bevilacqua, A., Di Giacomo, M., Remotti, D., Picchio, M.C., Fidanza, V., Cooper, M.D., Croce, C.M., et al. (2002). TCL1 participates in early embryonic development and is overexpressed in human seminomas. *Proc Natl Acad Sci U S A* 99, pp. 11712–7.

Neaves, W.B. (1971). Intercellular bridges between follicle cells and oocyte in the lizard, Anolis carolinensis. *Anat Rec* 170, pp. 285–301.

Newport, J., and Kirschner, M. (1982a). A major developmental transition in early *Xenopus* embryos: I. Characterization and timing of cellular changes at the midblastula stage. *Cell* 30, pp. 675–86.

Newport, J., and Kirschner, M. (1982b). A major developmental transition in early *Xenopus* embryos: II. Control of the onset of transcription. *Cell* 30, pp. 687–96.

Ng, A., Uribe, R.A., Yieh, L., Nuckels, R., and Gross, J.M. (2009). Zebrafish mutations in gart and paics identify crucial roles for de novo purine synthesis in vertebrate pigmentation and ocular development. *Development* 136, pp. 2601–11.

Nichols, J., Zevnik, B., Anastassiadis, K., Niwa, H., Klewe-Nebenius, D., Chambers, I., Scholer, H., and Smith, A. (1998). Formation of pluripotent stem cells in the mammalian embryo depends on the POU transcription factor Oct4. *Cell* 95, pp. 379–91.

Nicosia, S.V., Wolf, D.P., and Inoue, M. (1977). Cortical granule distribution and cell surface characteristics in mouse eggs. *Dev Biol* 57, pp. 56–74.

Niki, Y., and Mahowald, A.P. (2003). Ovarian cystocytes can repopulate the embryonic germ line and produce functional gametes. *Proc Natl Acad Sci U S A* 100, pp. 14042–5.

Nishioka, N., Inoue, K., Adachi, K., Kiyonari, H., Ota, M., Ralston, A., Yabuta, N., Hirahara, S., Stephenson, R.O., Ogonuki, N., et al. (2009). The Hippo signaling pathway components Lats and Yap pattern Tead4 activity to distinguish mouse trophectoderm from inner cell mass. *Dev Cell* 16, pp. 398–410.

Nishioka, N., Yamamoto, S., Kiyonari, H., Sato, H., Sawada, A., Ota, M., Nakao, K., and Sasaki, H. (2008). Tead4 is required for specification of trophectoderm in pre-implantation mouse embryos. *Mech Dev* 125, pp. 270–283.

Nojima, H., Shimizu, T., Kim, C.H., Yabe, T., Bae, Y.K., Muraoka, O., Hirata, T., Chitnis, A., Hirano, T., and Hibi, M. (2004). Genetic evidence for involvement of maternally derived Wnt canonical signaling in dorsal determination in zebrafish. *Mech Dev* 121, pp. 371–86.

Nojima, H., Rothhamel, S., Shimizu, T., Kim, C.H., Yonemura, S., Marlow, F.L., and Hibi, M. (2010). Syntabulin, a motor protein linker, controls dorsal determination. *Development* 137, pp. 923–33.

Nothias, J.Y., Majumder, S., Kaneko, K.J., and DePamphilis, M.L. (1995). Regulation of gene expression at the beginning of mammalian development. *J Biol Chem* 270, pp. 22077–80.

Nusse, R. (2005). Cell biology: relays at the membrane. *Nature* 438, pp. 747–9.

Nusslein-Volhard, C., and Wieschaus, E. (1980). Mutations affecting segment number and polarity in *Drosophila. Nature* 287, pp. 795–801.

Oh, B., Hwang, S.Y., Solter, D., and Knowles, B.B. (1997). Spindlin, a major maternal transcript expressed in the mouse during the transition from oocyte to embryo. *Development* 124, pp. 493–503.

Ohlstein, B., Lavoie, C.A., Vef, O., Gateff, E., and McKearin, D.M. (2000). The *Drosophila* cystoblast differentiation factor, benign gonial cell neoplasm, is related to DExH-box proteins and interacts genetically with bag-of-marbles. *Genetics* 155, pp. 1809–19.

Ohlstein, B., and McKearin, D. (1997). Ectopic expression of the *Drosophila* Bam protein eliminates oogenic germline stem cells. *Development* 124, pp. 3651–62.

Ohsugi, M., Zheng, P., Baibakov, B., Li, L., and Dean, J. (2008). Maternally derived FILIA-MATER complex localizes asymmetrically in cleavage-stage mouse embryos. *Development* 135, pp. 259–69.

Ohsumi, K., and Katagiri, C. (1991). Characterization of the ooplasmic factor inducing decondensation of and protamine removal from toad sperm nuclei: involvement of nucleoplasmin. *Dev Biol* 148, pp. 295–305.

Okuda, Y., Ogura, E., Kondoh, H., and Kamachi, Y. (2010). B1 SOX coordinate cell specification with patterning and morphogenesis in the early zebrafish embryo. *PLoS Genet* 6, p. e1000936.

Page, S.L., McKim, K.S., Deneen, B., Van Hook, T.L., and Hawley, R.S. (2000). Genetic studies of mei-P26 reveal a link between the processes that control germ cell proliferation in both sexes and those that control meiotic exchange in *Drosophila. Genetics* 155, pp. 1757–72.

Pan, G., Tian, S., Nie, J., Yang, C., Ruotti, V., Wei, H., Jonsdottir, G.A., Stewart, R., and Thomson, J.A. (2007). Whole-genome analysis of histone H3 lysine 4 and lysine 27 methylation in human embryonic stem cells. *Cell Stem Cell* 1, pp. 299–312.

Pant, V., Mariano, P., Kanduri, C., Mattsson, A., Lobanenkov, V., Heuchel, R., and Ohlsson, R. (2003). The nucleotides responsible for the direct physical contact between the chromatin insulator protein CTCF and the H19 imprinting control region manifest parent of origin-specific long-distance insulation and methylation-free domains. *Genes Dev* 17, pp. 586–90.

Parfenov, V., Potchukalina, G., Dudina, L., Kostyuchek, D., and Gruzova, M. (1989). Human antral follicles: oocyte nucleus and the karyosphere formation (electron microscopic and autoradiographic data). *Gamete Res* 22, pp. 219–31.

Parfenov, V.N., Pochukalina, G.N., Davis, D.S., Reinbold, R., Scholer, H.R., and Murti, K.G. (2003). Nuclear distribution of Oct-4 transcription factor in transcriptionally active and inactive mouse oocytes and its relation to RNA polymerase II and splicing factors. *J Cell Biochem* 89, pp. 720–32.

Paria, B.C., Dey, S.K., and Andrews, G.K. (1992). Antisense c-myc effects on preimplantation mouse embryo development. *Proc Natl Acad Sci U S A* 89, pp. 10051–55.

Parisi, M.J., Deng, W., Wang, Z., and Lin, H. (2001). The arrest gene is required for germline cyst formation during *Drosophila* oogenesis. *Genesis* 29, pp. 196–209.

Park, E.S., Lind, A.K., Dahm-Kahler, P., Brannstrom, M., Carletti, M.Z., Christenson, L.K., Curry, T.E., Jr., and Jo, M. (2010). RUNX2 Transcription Factor Regulates Gene Expression in Luteinizing Granulosa Cells of Rat Ovaries. *Mol Endocrinol* 24, pp. 846–58.

Pauken, C.M., and Capco, D.G. (1999). Regulation of cell adhesion during embryonic compaction of mammalian embryos: roles for PKC and beta-catenin. *Mol Reprod Dev* 54, pp. 135–44.

Payer, B., Saitou, M., Barton, S.C., Thresher, R., Dixon, J.P., Zahn, D., Colledge, W.H., Carlton, M.B., Nakano, T., and Surani, M.A. (2003). Stella is a maternal effect gene required for normal early development in mice. *Curr Biol* 13, pp. 2110–7.

Paynton, B.V., and Bachvarova, R. (1994). Polyadenylation and deadenylation of maternal mRNAs during oocyte growth and maturation in the mouse. *Mol Reprod Dev* 37, pp. 172–80.

Pei, W., and Feldman, B. (2009). Identification of common and unique modifiers of zebrafish midline bifurcation and cyclopia. *Dev Biol* 326, pp. 201–11.

Pei, W., Noushmehr, H., Costa, J., Ouspenskaia, M.V., Elkahloun, A.G., and Feldman, B. (2007a). An early requirement for maternal FoxH1 during zebrafish gastrulation. *Dev Biol* 310, pp. 10–22.

Pei, W., Williams, P.H., Clark, M.D., Stemple, D.L., and Feldman, B. (2007b). Environmental and genetic modifiers of squint penetrance during zebrafish embryogenesis. *Dev Biol* 308, pp. 368–78.

Pelegri, F. (2003). Maternal factors in zebrafish development. *Dev Dyn* 228, pp. 535–54.

Pelegri, F., and Mullins, M.C. (2004). Genetic screens for maternal-effect mutations. *Methods Cell Biol* 77, pp. 21–51.

Pelegri, F., Dekens, M.P., Schulte-Merker, S., Maischein, H.M., Weiler, C., and Nusslein-Volhard, C. (2004). Identification of recessive maternal-effect mutations in the zebrafish using a gynogenesis-based method. *Dev Dyn* 231, pp. 324–35.

Pelegri, F., Knaut, H., Maischein, H.M., Schulte-Merker, S., and Nusslein-Volhard, C. (1999). A mutation in the zebrafish maternal-effect gene nebel affects furrow formation and vasa RNA localization. *Curr Biol* 9, pp. 1431–40.

Pepling, M.E., de Cuevas, M., and Spradling, A.C. (1999). Germline cysts: a conserved phase of germ cell development? *Trends Cell Biol* 9, pp. 257–62.

Pepling, M.E., and Spradling, A.C. (1998). Female mouse germ cells form synchronously dividing cysts. *Development* 125, pp. 3323–8.

Pepling, M.E., and Spradling, A.C. (2001). Mouse ovarian germ cell cysts undergo programmed breakdown to form primordial follicles. *Dev Biol* 234, pp. 339–51.

Pepling, M.E., Wilhelm, J.E., O'Hara, A.L., Gephardt, G.W., and Spradling, A.C. (2007). Mouse oocytes within germ cell cysts and primordial follicles contain a Balbiani body. *Proc Natl Acad Sci U S A* 104, pp. 187–92.

Pesce, M., and Scholer, H.R. (2000). Oct-4: control of totipotency and germline determination. *Mol Reprod Dev* 55, pp. 452–57.

Peters, A.H., Kubicek, S., Mechtler, K., O'Sullivan, R.J., Derijck, A.A., Perez-Burgos, L., Kohlmaier, A., Opravil, S., Tachibana, M., Shinkai, Y., et al. (2003). Partitioning and plasticity of repressive histone methylation states in mammalian chromatin. *Mol Cell* 12, pp. 1577–89.

Pfannenstiel, H.D., and Grunig, C. (1982). Yolk formation in an annelid (Ophryotrocha puerilis, polychaeta). *Tissue Cell* 14, pp. 669–80.

Philipps, D.L., Wigglesworth, K., Hartford, S.A., Sun, F., Pattabiraman, S., Schimenti, K., Handel, M., Eppig, J.J., and Schimenti, J.C. (2008). The dual bromodomain and WD repeat-containing mouse protein BRWD1 is required for normal spermiogenesis and the oocyte-embryo transition. *Dev Biol* 317, pp. 72–82.

Philpott, A., Leno, G.H., and Laskey, R.A. (1991). Sperm decondensation in *Xenopus* egg cytoplasm is mediated by nucleoplasmin. *Cell* 65, pp. 569–78.

Piccolo, S., Agius, E., Leyns, L., Bhattacharyya, S., Grunz, H., Bouwmeester, T., and De Robertis, E.M. (1999). The head inducer Cerberus is a multifunctional antagonist of Nodal, BMP and Wnt signals. *Nature* 397, pp. 707–10.

Piko, L., and Clegg, K.B. (1982). Quantitative changes in total RNA, total poly(A), and ribosomes in early mouse embryos. *Dev Biol* 89, pp. 362–78.

Piotrowska, K., and Zernicka-Goetz, M. (2001). Role for sperm in spatial patterning of the early mouse embryo. *Nature* 409, pp. 517–21.

Plaster, N., Sonntag, C., Busse, C.E., and Hammerschmidt, M. (2006). p53 deficiency rescues apoptosis and differentiation of multiple cell types in zebrafish flathead mutants deficient for zygotic DNA polymerase delta1. *Cell Death Differ* 13, pp. 223–35.

Plusa, B., Hadjantonakis, A.K., Gray, D., Piotrowska-Nitsche, K., Jedrusik, A., Papaioannou, V.E.,

Glover, D.M., and Zernicka-Goetz, M. (2005). The first cleavage of the mouse zygote predicts the blastocyst axis. *Nature* 434, pp. 391–95.

Plusa, B., Piotrowska, K., and Zernicka-Goetz, M. (2002). Sperm entry position provides a surface marker for the first cleavage plane of the mouse zygote. *Genesis* 32, pp. 193–98.

Pogoda, H., Solnica-Krezel, L., Driever, W., and Meyer, D. (2000). The zebrafish forkhead transcription factor FoxH1/Fast1 is a modulator of nodal signaling required for organizer formation [In Process Citation]. *Curr Biol* 10, pp. 1041–49.

Poss, K.D. (2004). A zebrafish model of germ cell aneuploidy. *Cell Cycle* 3, pp. 1225–26.

Poss, K.D., Nechiporuk, A., Stringer, K.F., Lee, C., and Keating, M.T. (2004). Germ cell aneuploidy in zebrafish with mutations in the mitotic checkpoint gene mps1. *Genes Dev* 18, pp. 1527–32.

Ralston, A., and Rossant, J. (2008). Cdx2 acts downstream of cell polarization to cell-autonomously promote trophectoderm fate in the early mouse embryo. *Dev Biol* 313, pp. 614–29.

Ram, P.T., and Schultz, R.M. (1993). Reporter gene expression in G2 of the 1-cell mouse embryo. *Dev Biol* 156, pp. 552–56.

Ramos, S.B., Stumpo, D.J., Kennington, E.A., Phillips, R.S., Bock, C.B., Ribeiro-Neto, F., and Blackshear, P.J. (2004). The CCCH tandem zinc-finger protein Zfp36l2 is crucial for female fertility and early embryonic development. *Development* 131, pp. 4883–93.

Randall, R.A., Germain, S., Inman, G.J., Bates, P.A., and Hill, C.S. (2002). Different Smad2 partners bind a common hydrophobic pocket in Smad2 via a defined proline-rich motif. *EMBO J* 21, pp. 145–56.

Rane, M.J., Coxon, P.Y., Powell, D.W., Webster, R., Klein, J.B., Pierce, W., Ping, P., and McLeish, K.R. (2001). p38 Kinase-dependent MAPKAPK-2 activation functions as 3-phosphoinositide-dependent kinase-2 for Akt in human neutrophils. *J Biol Chem* 276, pp. 3517–23.

Rauch, G.J., Hammerschmidt, M., Blader, P., Schauerte, H.E., Strahle, U., Ingham, P.W., McMahon, A.P., and Haffter, P. (1997). Wnt5 is required for tail formation in the zebrafish embryo. *Cold Spring Harb Symp Quant Biol* 62, pp. 227–34.

Reim, G., Mizoguchi, T., Stainier, D.Y., Kikuchi, Y., and Brand, M. (2004). The POU domain protein spg (pou2/Oct4) is essential for endoderm formation in cooperation with the HMG domain protein casanova. *Dev Cell* 6, pp. 91–101.

Reim, G., and Brand, M. (2006). Maternal control of vertebrate dorsoventral axis formation and epiboly by the POU domain protein Spg/Pou2/Oct4. *Development* 133, pp. 2757–70.

Ren, Y., Cowan, R.G., Harman, R.M., and Quirk, S.M. (2009). Dominant activation of the hedgehog signaling pathway in the ovary alters theca development and prevents ovulation. *Mol Endocrinol* 23, pp. 711–23.

Rice, J.C., Briggs, S.D., Ueberheide, B., Barber, C.M., Shabanowitz, J., Hunt, D.F., Shinkai, Y., and Allis, C.D. (2003). Histone methyltransferases direct different degrees of methylation to define distinct chromatin domains. *Mol Cell* 12, pp. 1591–8.

Richardson, R.J., Dixon, J., Malhotra, S., Hardman, M.J., Knowles, L., Boot-Handford, R.P., Shore, P., Whitmarsh, A., and Dixon, M.J. (2006). Irf6 is a key determinant of the keratinocyte proliferation-differentiation switch. *Nat Genet* 38, pp. 1329–34.

Riparbelli, M.G., Massarelli, C., Robbins, L.G., and Callaini, G. (2004). The abnormal spindle protein is required for germ cell mitosis and oocyte differentiation during *Drosophila* oogenesis. *Exp Cell Res* 298, pp. 96–106.

Ro, H., and Dawid, I.B. (2009). Organizer restriction through modulation of Bozozok stability by the E3 ubiquitin ligase Lnx-like. *Nat Cell Biol* 11, pp. 1121–7.

Robinson, D.N., Smith-Leiker, T.A., Sokol, N.S., Hudson, A.M., and Cooley, L. (1997). Formation of the *Drosophila* ovarian ring canal inner rim depends on cheerio. *Genetics* 145, pp. 1063–72.

Roest, H.P., van Klaveren, J., de Wit, J., van Gurp, C.G., Koken, M.H., Vermey, M., van Roijen, J.H., Hoogerbrugge, J.W., Vreeburg, J.T., Baarends, W.M., et al. (1996). Inactivation of the HR6B ubiquitin-conjugating DNA repair enzyme in mice causes male sterility associated with chromatin modification. *Cell* 86, pp. 799–810.

Roest, H.P., Baarends, W.M., de Wit, J., van Klaveren, J.W., Wassenaar, E., Hoogerbrugge, J.W., van Cappellen, W.A., Hoeijmakers, J.H., and Grootegoed, J.A. (2004). The ubiquitin-conjugating DNA repair enzyme HR6A is a maternal factor essential for early embryonic development in mice. *Mol Cell Biol* 24, pp. 5485–95.

Rongo, C., Broihier, H.T., Moore, L., Van Doren, M., Forbes, A., and Lehmann, R. (1997). Germ plasm assembly and germ cell migration in *Drosophila*. *Cold Spring Harb Symp Quant Biol* 62, pp. 1–11.

Roper, K. (2007). Rtnl1 is enriched in a specialized germline ER that associates with ribonucleoprotein granule components. *J Cell Sci* 120, pp. 1081–92.

Roper, K., and Brown, N.H. (2004). A spectraplakin is enriched on the fusome and organizes microtubules during oocyte specification in *Drosophila*. *Curr Biol* 14, pp. 99–110.

Rosner, M.H., Vigano, M.A., Ozato, K., Timmons, P.M., Poirier, F., Rigby, P.W., and Staudt, L.M. (1990). A POU-domain transcription factor in early stem cells and germ cells of the mammalian embryo. *Nature* 345, pp. 686–92.

Roszko, I., Sawada, A., and Solnica-Krezel, L. (2009). Regulation of convergence and extension movements during vertebrate gastrulation by the Wnt/PCP pathway. Semin Cell *Dev Biol* 20, pp. 986–97.

Roulier, E.M., Panzer, S., and Beckendorf, S.K. (1998). The Tec29 tyrosine kinase is required during *Drosophila* embryogenesis and interacts with Src64 in ring canal development. *Mol Cell* 1, pp. 819–29.

Rowning, B.A., Wells, J., Wu, M., Gerhard, J.C., Moon, R.T., and Larabell, C.A. (1997). Microtubule-mediated transport of organells and localization of b-catenin to the future dorsal side of *Xenopus* eggs. *Proc Natl Acad Sci U S A* 94, pp. 1224–1229.

Ruchaud, S., Carmena, M., and Earnshaw, W.C. (2007). Chromosomal passengers: conducting cell division. *Nat Rev Mol Cell Biol* 8, pp. 798–812.

Rudolph, T., Yonezawa, M., Lein, S., Heidrich, K., Kubicek, S., Schafer, C., Phalke, S., Walther, M., Schmidt, A., Jenuwein, T., et al. (2007). Heterochromatin formation in *Drosophila* is initiated through active removal of H3K4 methylation by the LSD1 homolog SU(VAR)3-3. *Mol Cell* 26, pp. 103–15.

Ruggiu, M., Speed, R., Taggart, M., McKay, S.J., Kilanowski, F., Saunders, P., Dorin, J., and Cooke, H.J. (1997). The mouse Dazla gene encodes a cytoplasmic protein essential for gametogenesis. *Nature* 389, pp. 73–77.

Ruvinsky, I., Silver, L.M., and Ho, R.K. (1998). Characterization of the zebrafish tbx16 gene and evolution of the vertebrate T-box family. *Dev Genes Evol* 208, pp. 94–99.

Ryu, S., Holzschuh, J., Erhardt, S., Ettl, A.K., and Driever, W. (2005). Depletion of minichromosome maintenance protein 5 in the zebrafish retina causes cellcycle defect and apoptosis. *Proc Natl Acad Sci U S A* 102, pp. 18467–72.

Ryu, S.L., Fujii, R., Yamanaka, Y., Shimizu, T., Yabe, T., Hirata, T., Hibi, M., and Hirano, T. (2001). Regulation of dharma/bozozok by the Wnt pathway. *Dev Biol* 231, pp. 397–409.

Sabel, J.L., d'Alencon, C., O'Brien, E.K., Van Otterloo, E., Lutz, K., Cuykendall, T.N., Schutte, B.C., Houston, D.W., and Cornell, R.A. (2009). Maternal Interferon Regulatory Factor 6 is required for the differentiation of primary superficial epithelia in Danio and Xenopus embryos. *Dev Biol* 325, pp. 249–62.

Saito, D., Morinaga, C., Aoki, Y., Nakamura, S., Mitani, H., Furutani-Seiki, M., Kondoh, H., and Tanaka, M. (2007). Proliferation of germ cells during gonadal sex differentiation in medaka: Insights from germ cell-depleted mutant zenzai. *Dev Biol* 310, pp. 280–90.

Sanchez, R., and Marzluff, W.F. (2004). The oligo(A) tail on histone mRNA plays an active role in translational silencing of histone mRNA during *Xenopus* oogenesis. *Mol Cell Biol* 24, pp. 2513–25.

Santos, F., Hendrich, B., Reik, W., and Dean, W. (2002). Dynamic reprogramming of DNA methylation in the early mouse embryo. *Dev Biol* 241, pp. 172–82.

Santos-Rosa, H., Schneider, R., Bannister, A.J., Sherriff, J., Bernstein, B.E., Emre, N.C., Schreiber, S.L., Mellor, J., and Kouzarides, T. (2002). Active genes are tri-methylated at K4 of histone H3. *Nature* 419, pp. 407–11.

Saunders, P.T., Turner, J.M., Ruggiu, M., Taggart, M., Burgoyne, P.S., Elliott, D., and Cooke, H.J. (2003). Absence of mDazl produces a final block on germ cell development at meiosis. *Reproduction* 126, pp. 589–97.

Sawin, K.E., and Mitchison, T.J. (1995). Mutations in the kinesin-like protein Eg5 disrupting localization to the mitotic spindle. *Proc Natl Acad Sci U S A* 92, pp. 4289–3.

Schaner, C.E., Deshpande, G., Schedl, P.D., and Kelly, W.G. (2003). A conserved chromatin architecture marks and maintains the restricted germ cell lineage in worms and flies. *Dev Cell* 5, pp. 747–57.

Schier, A.F., and Giraldez, A.J. (2006). MicroRNA function and mechanism: insights from zebra fish. *Cold Spring Harb Symp Quant Biol* 71, pp. 195–203.

Schier, A.F., and Shen, M.M. (2000). Nodal signalling in vertebrate development. *Nature* 403, pp. 385–9.

Schier, A.F., Neuhauss, S.C.F., Harvey, M., Malicki, J., Solnica-Krezel, L., Stainier, D.Y.R., Zwartkruis, F., Abdelilah, S., Stemple, D.L., Rangini, Z., et al. (1996). Mutations affecting the development of the embryonic zebrafish brain. *Development, this issue.*

Schier, A.F., and Talbot, W.S. (2005). Molecular genetics of axis formation in zebrafish. *Annu Rev Genet* 39, pp. 561–613.

Schmid, B., Furthauer, M., Connors, S.A., Trout, J., Thisse, B., Thisse, C., and Mullins, M.C. (2000). Equivalent genetic roles for bmp7/snailhouse and bmp2b/swirl in dorsoventral pattern formation. *Development* 127, pp. 957–67.

Schneider, S., Steinbesser, H., Warga, R.M., and Hausen, P. (1996). β-catenin translocation into nuclei demarcates the dorsalizing centers in frog and fish embryos. Mechanisms of *Development* 57, pp. 191–98.

Scholer, H.R., Ruppert, S., Suzuki, N., Chowdhury, K., and Gruss, P. (1990). New type of POU domain in germ line-specific protein Oct-4. *Nature* 344, pp. 435–39.

Schramm, R.D., Tennier, M.T., Boatman, D.E., and Bavister, B.D. (1993). Chromatin configurations and meiotic competence of oocytes are related to follicular diameter in nonstimulated rhesus monkeys. *Biol Reprod* 48, pp. 349–56.

Schubeler, D., MacAlpine, D.M., Scalzo, D., Wirbelauer, C., Kooperberg, C., van Leeuwen, F., Gottschling, D.E., O'Neill, L.P., Turner, B.M., Delrow, J., et al. (2004). The histone modification pattern of active genes revealed through genome-wide chromatin analysis of a higher eukaryote. *Genes Dev* 18, pp. 1263–71.

Schultz, G.A., and Heyner, S. (1992). Gene expressin in pre-implantation mammalian embryos. *Mutation Research* 296, pp. 17–31.

Schulte-Merker, S., VanEeden, F.J.M., Halpern, M.E., Kimmel, C.B., and Nusslein-Volhard, C. (1994). No tail (Ntl) is the zebrafish homologue of the mouse T (Brachyury) gene. *Development* 120, pp. 1009–15.

Schulz, C., Kiger, A.A., Tazuke, S.I., Yamashita, Y.M., Pantalena-Filho, L.C., Jones, D.L., Wood, C.G., and Fuller, M.T. (2004). A misexpression screen reveals effects of bag-of-marbles and TGF beta class signaling on the Drosophila male germline stem cell lineage. *Genetics* 167, pp. 707–23.

Schultz, R.M. (1993). Regulation of zygotic gene activation in the mouse. *Bioessays* 15, pp. 531–38.

Schultz, R.M. (2002). The molecular foundations of the maternal to zygotic transition in the preimplantation embryo. *Hum Reprod Update* 8, pp. 323–31.

Selman, K., Wallace, R., Sarka, A., and Qi, X. (1993). Stages of oocyte development in the zebrafish, *Brachydanio rerio. J Morphol* 218, pp. 203–24.

Seto, A.G., Kingston, R.E., and Lau, N.C. (2007). The coming of age for Piwi proteins. *Mol Cell* 26, pp. 603–9.

Shimizu, T., Yabe, T., Muraoka, O., Yonemura, S., Aramaki, S., Hatta, K., Bae, Y.K., Nojima, H., and Hibi, M. (2005). E-cadherin is required for gastrulation cell movements in zebrafish. *Mech Dev* 122, pp. 747–63.

Shivdasani, A.A., and Ingham, P.W. (2003). Regulation of stem cell maintenance and transit amplifying cell proliferation by tgf-beta signaling in *Drosophila* spermatogenesis. *Curr Biol* 13, pp. 2065–72.

Sidi, S., Goutel, C., Peyrieras, N., and Rosa, F.M. (2003). Maternal induction of ventral fate by zebrafish radar. *Proc Natl Acad Sci U S A* 100, pp. 3315–20.

Sil, A.K., Maeda, S., Sano, Y., Roop, D.R., and Karin, M. (2004). IkappaB kinase-alpha acts in the epidermis to control skeletal and craniofacial morphogenesis. *Nature* 428, pp. 660–64.

Sirotkin, H.I., Gates, M.A., Kelly, P.D., Schier, A.F., and Talbot, W.S. (2000). Fast1 is required for the development of dorsal axial structures in zebrafish. *Curr Biol* 10, pp. 1051–54.

Sirotkin, H.I., Dougan, S.T., Schier, A.F., and Talbot, W.S. (2000a). Bozozok and squint act in parallel to specify dorsal mesoderm and anterior neuroectoderm in zebrafish. *Development* 127, pp. 2583–92.

Sirotkin, H.I., Gates, M.A., Kelly, P.D., Schier, A.F., and Talbot, W.S. (2000b). Fast1 is required for the development of dorsal axial structures in zebrafish. *Curr Biol* 10, pp. 1051–54.

Slanchev, K., Carney, T.J., Stemmler, M.P., Koschorz, B., Amsterdam, A., Schwarz, H., and Hammerschmidt, M. (2009). The epithelial cell adhesion molecule EpCAM is required for epithelial morphogenesis and integrity during zebrafish epiboly and skin development. *PLoS Genet* 5, p. e1000563.

Slanchev, K., Stebler, J., de la Cueva-Mendez, G., and Raz, E. (2005). *Development* without germ cells: the role of the germ line in zebrafish sex differentiation. *Proc Natl Acad Sci U S A* 102, pp. 4074–79.

Snapp, E.L., Iida, T., Frescas, D., Lippincott-Schwartz, J., and Lilly, M.A. (2004). The fusome mediates intercellular endoplasmic reticulum connectivity in *Drosophila* ovarian cysts. *Mol Biol Cell* 15, pp. 4512–21.

Sokol, S., Christian, J.L., Moon, R.T., and Melton, D.A. (1991). Injected Wnt RNA induces a complete body axis in *Xenopus* embryos. *Cell* 67, pp. 741–52.

Solnica-Krezel, L. (1999). Pattern formation in zebrafish—fruitful liasons between embryology and genetics. *Curr Top Dev Biol* 41, pp. 1–35.

Solnica-Krezel, L., Schier, A.F., and Driever, W. (1994). Efficient recovery of ENU-induced mutations from the zebrafish germline. *Genetics* 136, pp. 1401–20.

Solnica-Krezel, L. (2006). Gastrulation in zebrafish—all just about adhesion? *Curr Opin Genet Dev* 16, pp. 433–41.

Solnica-Krezel L., and Driever W. (1994). Microtubule arrays of the zebrafish yolk cell: organization and function during epiboly. *Development* 120, pp. 2443–55.

Soyal, S.M., Amleh, A., and Dean, J. (2000). FIGalpha, a germ cell-specific transcription factor required for ovarian follicle formation. *Development* 127, pp. 4645–54.

St Johnston, D., and Nusslein-Volhard, C. (1992). The origin of pattern and polarity in the *Drosophila* embryo. *Cell* 68, pp. 201–19.

Stachel, S.E., Grunwald, D.J., and Myers, P.Z. (1993). Lithium perturbation and goosecoid expression identify a dorsal specification pathway in the pregastrula zebrafish. *Development* 117, pp. 1261–74.

Stancheva, I., and Meehan, R.R. (2000). Transient depletion of xDnmt1 leads to premature gene activation in *Xenopus* embryos. *Genes Dev* 14, pp. 313–27.

Standley, H.J., Destree, O., Kofron, M., Wylie, C., and Heasman, J. (2006). Maternal XTcf1 and XTcf4 have distinct roles in regulating Wnt target genes. *Dev Biol* 289, pp. 318–28.

Storto, P.D., and King, R.C. (1989). The role of polyfusomes in generating branched chains of cystocytes during *Drosophila* oogenesis. *Dev Genet* 10, pp. 70–86.

Strahl, B.D., and Allis, C.D. (2000). The language of covalent histone modifications. *Nature* 403, pp. 41–45.

Strahle, U., and Jesuthasan, S. (1993). Ultraviolet irradiation impairs epiboly in zebrafish embryos: evidence for a microtubule-dependent mechanism of epiboly. *Development* 119, pp. 909–19.

Strumpf, D., Mao, C.A., Yamanaka, Y., Ralston, A., Chawengsaksophak, K., Beck, F., and Rossant, J. (2005). Cdx2 is required for correct cell fate specification and differentiation of trophectoderm in the mouse blastocyst. *Development* 132, pp. 2093–102.

Su, Y.Q., Wu, X., O'Brien, M.J., Pendola, F.L., Denegre, J.N., Matzuk, M.M., and Eppig, J.J. (2004). Synergistic roles of BMP15 and GDF9 in the development and function of the oocyte-cumulus cell complex in mice: genetic evidence for an oocyte-granulosa cell regulatory loop. *Dev Biol* 276, pp. 64–73.

Su, Y.Q., Sugiura, K., Li, Q., Wigglesworth, K., Matzuk, M.M., and Eppig, J.J. (2010). Mouse oocytes enable LH-induced maturation of the cumulus-oocyte complex via promoting EGF receptor-dependent signaling. *Mol Endocrinol* 24, pp. 1230–39.

Suh, N., Baehner, L., Moltzahn, F., Melton, C., Shenoy, A., Chen, J., and Blelloch, R. (2010).

MicroRNA function is globally suppressed in mouse oocytes and early embryos. *Curr Biol* 20, pp. 271–77.

Sullivan, E., Santiago, C., Parker, E.D., Dominski, Z., Yang, X., Lanzotti, D.J., Ingledue, T.C., Marzluff, W.F., and Duronio, R.J. (2001). *Drosophila* stem loop binding protein coordinates accumulation of mature histone mRNA with cell cycle progression. *Genes Dev* 15, pp. 173–87.

Sumanas, S., Strege, P., Heasman, J., and Ekker, S.C. (2000). The putative wnt receptor *Xenopus* frizzled-7 functions upstream of beta-catenin in vertebrate dorsoventral mesoderm patterning. *Development* 127, pp. 1981–90.

Sumanas, S., and Ekker, S.C. (2001). *Xenopus* frizzled-7 morphant displays defects in dorsoventral patterning and convergent extension movements during gastrulation. *Genesis* 30, pp. 119–22.

Sundaram, N., Tao, Q., Wylie, C., and Heasman, J. (2003). The role of maternal CREB in early embryogenesis of *Xenopus laevis*. *Dev Biol* 261, pp. 337–52.

Suter, B., Romberg, L.M., and Steward, R. (1989). Bicaudal-D, a *Drosophila* gene involved in developmental asymmetry: localized transcript accumulation in ovaries and sequence similarity to myosin heavy chain tail domains. *Genes Dev* 3, pp. 1957–68.

Suter, B., and Steward, R. (1991). Requirement for phosphorylation and localization of the Bicaudal-D protein in *Drosophila* oocyte differentiation. *Cell* 67, pp. 917–26.

Svoboda, P., Stein, P., Hayashi, H., and Schultz, R.M. (2000). Selective reduction of dormant maternal mRNAs in mouse oocytes by RNA interference. *Development* 127, pp. 4147–56.

Svoboda, P., Stein, P., and Schultz, R.M. (2001). RNAi in mouse oocytes and preimplantation embryos: effectiveness of hairpin dsRNA. *Biochem Biophys Res Commun* 287, pp. 1099–104.

Szabo, P.E., Tang, S.H., Silva, F.J., Tsark, W.M., and Mann, J.R. (2004). Role of CTCF binding sites in the Igf2/H19 imprinting control region. *Mol Cell Biol* 24, pp. 4791–800.

Szakmary, A., Cox, D.N., Wang, Z., and Lin, H. (2005). Regulatory relationship among piwi, pumilio, and bag-of-marbles in Drosophila germline stem cell selfrenewal and differentiation. *Curr Biol* 15, pp. 171–8.

Tada, M., and Kai, M. (2009). Noncanonical Wnt/PCP signaling during vertebrate gastrulation. *Zebrafish* 6, pp. 29–40.

Takahashi, K., and Yamanaka, S. (2006). Induction of pluripotent stem cells from mouse embryonic and adult fibroblast cultures by defined factors. *Cell* 126, pp. 663–76.

Takeda, K., Takeuchi, O., Tsujimura, T., Itami, S., Adachi, O., Kawai, T., Sanjo, H., Yoshikawa, K., Terada, N., and Akira, S. (1999). Limb and skin abnormalities in mice lacking IKKalpha. *Science* 284, pp. 313–6.

Takeda, Y., Mishima, Y., Fujiwara, T., Sakamoto, H., and Inoue, K. (2009). DAZL relieves miRNA-mediated repression of germline mRNAs by controlling poly(A) tail length in zebrafish. *PLoS One* 4, p. e7513.

Tanegashima, K., Haramoto, Y., Yokota, C., Takahashi, S., and Asashima, M. (2004). Xantivin suppresses the activity of EGF-CFC genes to regulate nodal signaling. *Int J Dev Biol* 48, pp. 275–83.

Tan, C., Stronach, B., and Perrimon, N. (2003). Roles of myosin phosphatase during *Drosophila* development. *Development* 130, pp. 671–81.

Tannahill, D., and Melton, D.A. (1989). Localized synthesis of the Vg1 protein during early *Xenopus* development. *Development* 106, pp. 775–85.

Tao, Q., Nandadasa, S., McCrea, P.D., Heasman, J., and Wylie, C. (2007). Gprotein-coupled signals control cortical actin assembly by controlling cadherin expression in the early *Xenopus* embryo. *Development* 134, pp. 2651–61.

Tao, Q., Yokota, C., Puck, H., Kofron, M., Birsoy, B., Yan, D., Asashima, M., Wylie, C.C., Lin, X., and Heasman, J. (2005). Maternal wnt11 activates the canonical wnt signaling pathway required for axis formation in *Xenopus* embryos. *Cell* 120, pp. 857–71.

Tashiro, F., Kanai-Azuma, M., Miyazaki, S., Kato, M., Tanaka, T., Toyoda, S., Yamato, E., Kawa kami, H., Miyazaki, T., and Miyazaki, J. (2010). Maternal-effect gene Ces5/Ooep/Moep19/Floped is essential for oocyte cytoplasmic lattice formation and embryonic development at the maternal-zygotic stage transition. *Genes Cells* 15, pp. 813–28.

Temeles, G.L., Ram, P.T., Rothstein, J.L., and Schultz, R.M. (1994). Expression patterns of novel genes during mouse preimplantation embryogenesis. *Mol Reprod Dev* 37, pp. 121–9.

Theurkauf, W.E., Alberts, B.M., Jan, Y.N., and Jongens, T.A. (1993). A central role for microtubules in the differentiation of *Drosophila* oocytes. *Development* 118, pp. 1169–80.

Thompson, E.M., Legouy, E., and Renard, J.P. (1998). Mouse embryos do not wait for the MBT: chromatin and RNA polymerase remodeling in genome activation at the onset of development. *Dev Genet* 22, pp. 31–42.

Tian, Q., Kopf, G.S., Brown, R.S., and Tseng, H. (2001). Function of basonuclin in increasing transcription of the ribosomal RNA genes during mouse oogenesis. *Development* 128, pp. 407–16.

Tong, Z.B., and Nelson, L.M. (1999). A mouse gene encoding an oocyte antigen associated with autoimmune premature ovarian failure. *Endocrinology* 140, pp. 3720–6.

Tong, Z.B., Gold, L., Pfeifer, K.E., Dorward, H., Lee, E., Bondy, C.A., Dean, J., and Nelson, L.M. (2000a). Mater, a maternal effect gene required for early embryonic development in mice. *Nat Genet* 26, pp. 267–68.

Tong, Z.B., Nelson, L.M., and Dean, J. (2000b). Mater encodes a maternal protein in mice with a leucine-rich repeat domain homologous to porcine ribonuclease inhibitor. *Mamm Genome* 11, pp. 281–87.

Topczewski, J., Sepich, D.S., Myers, D.C., Walker, C., Amores, A., Lele, Z., Hammerschmidt, M., Postlethwait, J., and Solnica-Krezel, L. (2001). The zebrafish glypican knypek controls cell polarity during gastrulation movements of convergent extension. *Developmental Cell* 1, pp. 251–64.

Torres-Padilla, M.E., and Zernicka-Goetz, M. (2006). Role of TIF1alpha as a modulator of embryonic transcription in the mouse zygote. *J Cell Biol* 174, pp. 329–38.

Trentini, M., and Scanabissi, F.S. (1978). Ultrastructural observations on the oogenesis of Triops cancriformis (Crustacea, Notostraca). *Cell Tissue Res* 194, pp. 71–7.

Tsaadon, A., Eliyahu, E., Shtraizent, N., and Shalgi, R. (2006). When a sperm meets an egg: block to polyspermy. *Mol Cell Endocrinol* 252, pp. 107–14.

Tseng, H., Biegel, J.A., and Brown, R.S. (1999). Basonuclin is associated with the ribosomal RNA genes on human keratinocyte mitotic chromosomes. *J Cell Sci* 112 Pt 18, pp. 3039–47.

Tsujikawa, M., and Malicki, J. (2004). Intraflagellar transport genes are essential for differentiation and survival of vertebrate sensory neurons. *Neuron* 42, pp 703–16.

Tsukamoto, S., Kuma, A., and Mizushima, N. (2008a). The role of autophagy during the oocyte-to-embryo transition. *Autophagy* 4, pp. 1076–78.

Tsukamoto, S., Kuma, A., Murakami, M., Kishi, C., Yamamoto, A., and Mizushima, N. (2008b). Autophagy is essential for preimplantation development of mouse embryos. *Science* 321, pp. 117–20.

Turner, B.M. (2000). Histone acetylation and an epigenetic code. *Bioessays* 22, pp. 836–45.

Vaccari, T., and Ephrussi, A. (2002). The fusome and microtubules enrich Par-1 in the oocyte, where it effects polarization in conjunction with Par-3, BicD, Egl, and dynein. *Curr Biol* 12, pp. 1524–8.

Vastenhouw, N.L., Zhang, Y., Woods, I.G., Imam, F., Regev, A., Liu, X.S., Rinn, J., and Schier, A.F. (2010). Chromatin signature of embryonic pluripotency is established during genome activation. *Nature* 464, pp. 922–6.

Veenstra, G.J., Destree, O.H., and Wolffe, A.P. (1999). Translation of maternal TATA-binding protein mRNA potentiates basal but not activated transcription in *Xenopus* embryos at the midblastula transition. *Mol Cell Biol* 19, pp. 7972–82.

Veien, E.S., Grierson, M.J., Saund, R.S., and Dorsky, R.I. (2005). Expression pattern of zebrafish tcf7 suggests unexplored domains of Wnt/beta-catenin activity. *Dev Dyn* 233, pp. 233–9.

Vinot, S., Le, T., Maro, B., and Louvet-Vallee, S. (2004). Two PAR6 proteins become asymmetrically localized during establishment of polarity in mouse oocytes. *Curr Biol* 14, pp. 520–25.

Vinot, S., Le, T., Ohno, S., Pawson, T., Maro, B., and Louvet-Vallee, S. (2005). Asymmetric distribution of PAR proteins in the mouse embryo begins at the 8-cell stage during compaction. *Dev Biol* 282, pp. 307–19.

Wada, H., Iwasaki, M., Sato, T., Masai, I., Nishiwaki, Y., Tanaka, H., Sato, A., Nojima, Y., and Okamoto, H. (2005). Dual roles of zygotic and maternal Scribble1 in neural migration and convergent extension movements in zebrafish embryos. *Development* 132, pp. 2273–85.

Wadsworth, P., and Khodjakov, A. (2004). E pluribus unum: towards a universal mechanism for spindle assembly. *Trends Cell Biol* 14, pp. 413–19.

Wagner, D.S., Dosch, R., Mintzer, K.A., Wiemelt, A.P., and Mullins, M.C. (2004). Maternal control of development at the midblastula transition and beyond: mutants from the zebrafish II. *Dev Cell* 6, pp. 781–90.

Wallace, R.A., and Selman, K. (1990). Ultrastructural aspects of oogenesis and oocyte growth in fish and amphibians. *J Electron Microsc Tech* 16, pp. 175–201.

Wallingford, J.B. (2005). Vertebrate gastrulation: polarity genes control the matrix. *Curr Biol* 15, pp. R414–16.

Wan, L.B., Pan, H., Hannenhalli, S., Cheng, Y., Ma, J., Fedoriw, A., Lobanenkov, V., Latham, K.E., Schultz, R.M., and Bartolomei, M.S. (2008). Maternal depletion of CTCF reveals multiple functions during oocyte and preimplantation embryo development. *Development* 135, pp. 2729–38.

Wang, Z.F., Whitfield, M.L., Ingledue, T.C., 3rd, Dominski, Z., and Marzluff, W.F. (1996). The protein that binds the 3' end of histone mRNA: a novel RNA-binding protein required for histone pre-mRNA processing. *Genes Dev* 10, pp. 3028–40.

Wang, Y., Medvid, R., Melton, C., Jaenisch, R., and Blelloch, R. (2007). DGCR8 is essential for microRNA biogenesis and silencing of embryonic stem cell self-renewal. *Nat Genet* 39, pp. 380–85.

Warga, R.M., and Kimmel, C.B. (1990). Cell movements during epiboly and gastrulation in zebrafish. *Development* 108, pp. 569–80.

Wassarman, P.M., and Kinloch, R.A. (1992). Gene expression during oogenesis in mice. *Mutat Res* 296, pp. 3–15.

Weaver, C., Farr, G.H., 3rd, Pan, W., Rowning, B.A., Wang, J., Mao, J., Wu, D., Li, L., Larabell, C.A., and Kimelman, D. (2003). GBP binds kinesin light chain and translocates during cortical rotation in *Xenopus* eggs. *Development* 130, pp. 5425–36.

Webb, T.A., Kowalski, W.J., and Fluck, R.A. (1995). Microtubule-based movements during ooplasmic segregation in the medaka fish egg (*Oryzias latipes*). *Biological Bulletin* 188, pp. 146–56.

Wei, G., Oliver, B., Pauli, D., and Mahowald, A.P. (1994). Evidence for sex transformation of germline cells in ovarian tumor mutants of Drosophila. *Dev Biol* 161, pp. 318–20.

Weidinger, G., Stebler, J., Slanchev, K., Dumstrei, K., Wise, C., Lovell-Badge, R., Thisse, C., Thisse, B., and Raz, E. (2003). Dead end, a novel vertebrate germ plasm component, is required for zebrafish primordial germ cell migration and survival. *Curr Biol* 13, pp. 1429–34.

Westfall, T.A., Brimeyer, R., Twedt, J., Gladon, J., Olberding, A., Furutani-Seiki, M., and Slusarski, D.C. (2003). Wnt-5/pipetail functions in vertebrate axis formation as a negative regulator of Wnt/beta-catenin activity. *J Cell Biol* 162, pp. 889–98.

White, J.A., and Heasman, J. (2008). Maternal control of pattern formation in *Xenopus laevis*. *J Exp Zool B Mol Dev Evol* 310, pp. 73–84.

Whitfield, M.L., Zheng, L.X., Baldwin, A., Ohta, T., Hurt, M.M., and Marzluff, W.F. (2000). Stem-loop binding protein, the protein that binds the 3' end of histone mRNA, is cell cycle regulated by both translational and posttranslational mechanisms. *Mol Cell Biol* 20, pp. 4188–98.

Wiekowski, M., Miranda, M., and DePamphilis, M.L. (1993). Requirements for promoter activity in mouse oocytes and embryos distinguish paternal pronuclei from maternal and zygotic nuclei. *Dev Biol* 159, pp. 366–78.

Wiekowski, M., Miranda, M., Nothias, J.Y., and DePamphilis, M.L. (1997). Changes in histone synthesis and modification at the beginning of mouse development correlate with the establishment of chromatin mediated repression of transcription. *J Cell Sci* 110 (Pt 10), pp. 1147–58.

Wilding, M., Carotenuto, R., Infante, V., Dale, B., Marino, M., Di Matteo, L., and Campanella, C. (2001). Confocal microscopy analysis of the activity of mitochondria contained within the 'mitochondrial cloud' during oogenesis in *Xenopus laevis*. *Zygote* 9, pp. 347–52.

Wilhelm, J.E., Buszczak, M., and Sayles, S. (2005). Efficient protein trafficking requires trailer hitch, a component of a ribonucleoprotein complex localized to the ER in *Drosophila*. *Dev Cell* 9, pp. 675–85.

Wilk, K., Bilinski, S., Dougherty, M.T., and Kloc, M. (2005). Delivery of germinal granules and localized RNAs via the messenger transport organizer pathway to the vegetal cortex of *Xenopus* oocytes occurs through directional expansion of the mitochondrial cloud. *Int J Dev Biol* 49, pp. 17–21.

Willert, K., Shibamoto, S., and Nusse, R. (1999). Wnt-induced dephosphorylation of axin releases beta-catenin from the axin complex. *Genes Dev* 13, pp. 1768–73.

Wilmut, I., Schnieke, A.E., McWhir, J., Kind, A.J., and Campbell, K.H. (1997). Viable offspring derived from fetal and adult mammalian cells. *Nature* 385, pp. 810–13.

Winston, N., Bourgain-Guglielmetti, F., Ciemerych, M.A., Kubiak, J.Z., Senamaud-Beaufort, C., Carrington, M., Brechot, C., and Sobczak-Thepot, J. (2000). Early development of mouse embryos null mutant for the cyclin A2 gene occurs in the absence of maternally derived cyclin A2 gene products. *Dev Biol* 223, pp. 139–53.

Wiznerowicz, M., Jakobsson, J., Szulc, J., Liao, S., Quazzola, A., Beermann, F., Aebischer, P., and Trono, D. (2007). The Kruppel-associated box repressor domain can trigger de novo promoter methylation during mouse early embryogenesis. *J Biol Chem* 282, pp. 34535–41.

Wodarz, A. (2002). Establishing cell polarity in development. *Nat Cell Biol* 4, pp. E39–44.

Wong, J.L., and Wessel, G.M. (2006). Defending the zygote: search for the ancestral animal block to polyspermy. *Curr Top Dev Biol* 72, pp. 1–151.

Worrad, D.M., Ram, P.T., and Schultz, R.M. (1994). Regulation of gene expression in the mouse oocyte and early preimplantation embryo: developmental changes in Sp1 and TATA box-binding protein, TBP. *Development* 120, pp. 2347–57.

Wu, X., Wang, P., Brown, C.A., Zilinski, C.A., and Matzuk, M.M. (2003). Zygote arrest 1 (Zar1) is an evolutionarily conserved gene expressed in vertebrate ovaries. *Biol Reprod* 69, pp. 861–67.

Wu, Y., Zhan, L., Ai, Y., Hannigan, M., Gaestel, M., Huang, C.K., and Madri, J.A. (2007). MAPKAPK2-mediated LSP1 phosphorylation and FMLP-induced neutrophil polarization. *Biochem Biophys Res Commun* 358, pp. 170–75.

Wylie, C. (2000). Germ cells. *Curr Opin Genet Dev* 10, pp. 410–3.

Wylie, C., Kofron, M., Payne, C., Anderson, R., Hosobuchi, M., Joseph, E., and Heasman, J. (1996). Maternal beta-catenin establishes a 'dorsal signal' in early *Xenopus* embryos. *Development* 122, pp. 2987–96.

Wylie, C.C., Holwill, S., O'Driscoll, M., Snape, A., and Heasman, J. (1985). Germ plasm and germ cell determination in *Xenopus laevis* as studied by cell transplantation analysis. *Cold Spring Harb Symp Quant Biol* 50, pp. 37–43.

Xanthos, J.B., Kofron, M., Wylie, C., and Heasman, J. (2001). Maternal VegT is the initiator of a molecular network specifying endoderm in *Xenopus laevis*. *Development* 128, pp. 167–80.

Xi, R., Doan, C., Liu, D., and Xie, T. (2005). Pelota controls self-renewal of germline stem cells by repressing a Bam-independent differentiation pathway. *Development* 132, pp. 5365–74.

Yabe, T., Ge, X., Lindeman, R., Nair, S., Runke, G., Mullins, M.C., and Pelegri, F. (2009). The maternal-effect gene cellular island encodes aurora B kinase and is essential for furrow formation in the early zebrafish embryo. *PLoS Genet* 5, p. e1000518.

Yabe, T., Ge, X., and Pelegri, F. (2007). The zebrafish maternal-effect gene cellular atoll encodes the centriolar component sas-6 and defects in its paternal function promote whole genome duplication. *Dev Biol* (in press).

Yagi, R., Kohn, M.J., Karavanova, I., Kaneko, K.J., Vullhorst, D., DePamphilis, M.L., and Buonanno, A. (2007). Transcription factor TEAD4 specifies the trophectoderm lineage at the beginning of mammalian development. *Development* 134, pp. 3827–36.

Yamamoto, M., Meno, C., Sakai, Y., Shiratori, H., Mochida, K., Ikawa, Y., Saijoh, Y., and Hamada, H. (2001). The transcription factor FoxH1 (FAST) mediates Nodal signaling during anterior-posterior patterning and node formation in the mouse. *Genes Dev* 15, pp. 1242–56.

Yamashita, M., Fukada, S., Yoshikuni, M., Bulet, P., Hirai, T., Yamaguchi, A., Lou, Y.H., Zhao, Z., and Nagahama, Y. (1992). Purification and characterization of maturation-promoting factor in fish. *Dev Biol* 149, pp. 8–15.

Yan, C., Wang, P., DeMayo, J., DeMayo, F.J., Elvin, J.A., Carino, C., Prasad, S.V., Skinner, S.S., Dunbar, B.S., Dube, J.L., et al. (2001). Synergistic roles of bone morphogenetic protein 15 and growth differentiation factor 9 in ovarian function. *Mol Endocrinol* 15, pp. 854–66.

Yan, W., Rajkovic, A., Viveiros, M.M., Burns, K.H., Eppig, J.J., and Matzuk, M.M. (2002). Identification of Gasz, an evolutionarily conserved gene expressed exclusively in germ cells andencoding a protein with four ankyrin repeats, a sterile-alpha motif, and a basic leucine zipper. *Mol Endocrinol* 16, pp. 1168–84.

Yan, W., Ma, L., Zilinski, C.A., and Matzuk, M.M. (2004). Identification and characterization of evolutionarily conserved pufferfish, zebrafish, and frog orthologs of GASZ. *Biol Reprod* 70, pp. 1619–25.

Yan, X., Yu, S., Lei, A., Hua, J., Chen, F., Li, L., Xie, X., Yang, X., Geng, W., and Dou, Z. (2010). The four reprogramming factors and embryonic development in mice. *Cell Reprogram* 12, pp. 565–70.

Yang, J., Tan, C., Darken, R.S., Wilson, P.A., and Klein, P.S. (2002). Betacatenin/Tcf-regulated transcription prior to the midblastula transition. *Development* 129, pp. 5743–52.

Yanagimachi, R. (1994). Fertility of mammalian spermatozoa: its development and relativity. *Zygote* 2, pp. 371–72.

Yeo, C., and Whitman, M. (2001). Nodal signals to Smads through Criptodependent and Cripto-independent mechanisms. *Mol Cell* 7, pp. 949–57.

Yeom, Y.I., Fuhrmann, G., Ovitt, C.E., Brehm, A., Ohbo, K., Gross, M., Hubner, K., and Scholer, H.R. (1996). Germline regulatory element of Oct-4 specific for the totipotent cycle of embryonal cells. *Development* 122, pp. 881–94.

Yin, C., Ciruna, B., and Solnica-Krezel, L. (2009). Convergence and extension movements during vertebrate gastrulation. *Curr Top Dev Biol* 89, pp. 163–92.

Yisraeli, J.K., and Melton, D.A. (1988). The maternal mRNA Vg1 is correctly localized following injection into *Xenopus* oocytes. *Nature* 336, pp. 592–95.

Yisraeli, J.K., Sokol, S., and Melton, D.A. (1989). The process of localizing a maternal messenger RNA in *Xenopus* oocytes. *Development* 107 (Suppl), pp. 31–36.

Yoon, C., Kawakami, K., and Hopkins, N. (1997). Zebrafish vasa homologue RNA is localized to the cleavage planes of 2- and 4-cell-stage embryos and is expressed in the primordial germ cells. *Development* 124, pp. 3157–65.

Yost, C., Farr III, G.H., Pierce, S.B., Ferkey, D.M., Mingzi Chen, M., and Kimelman, D. (1998). GBP, an inhibitor of GSK-3, is implicated in *Xenopus* development and oncogenesis. *Cell* 93, pp. 1031–41.

Yuba-Kubo, A., Kubo, A., Hata, M., and Tsukita, S. (2005). Gene knockout analysis of two gamma-tubulin isoforms in mice. *Dev Biol* 282, pp. 361–73.

Yue, L., and Spradling, A.C. (1992). Hu-li tai shao, a gene required for ring canal formation during *Drosophila* oogenesis, encodes a homolog of adducin. *Genes Dev* 6, pp. 2443–54.

Yurttas, P., Vitale, A.M., Fitzhenry, R.J., Cohen-Gould, L., Wu, W., Gossen, J.A., and Coonrod,

S.A. (2008). Role for PADI6 and the cytoplasmic lattices in ribosomal storage in oocytes and translational control in the early mouse embryo. *Development* 135, pp. 2627–36.

Zalik, S.E., Lewandowski, E., Kam, Z., and Geiger, B. (1999). Cell adhesion and the actin cytoskeleton of the enveloping layer in the zebrafish embryo during epiboly. *Biochem Cell Biol* 77, pp. 527–42.

Zamir, E., Kam, Z., and Yarden, A. (1997). Transcription-dependent induction of G1 phase during the zebrafish midblastula transition. *Mol Cell Biol* 17, pp. 529–36.

Zeng, F., and Schultz, R.M. (2003). Gene expression in mouse oocytes and preimplantation embryos: use of suppression subtractive hybridization to identify oocyte- and embryo-specific genes. *Biol Reprod* 68, pp. 31–39.

Zernicka-Goetz, M. (2002). Patterning of the embryo: the first spatial decisions in the life of a mouse. *Development* 129, pp. 815–29.

Zernicka-Goetz, M. (2006). The first cell-fate decisions in the mouse embryo: destiny is a matter of both chance and choice. *Curr Opin Genet Dev* 16, pp. 406–12.

Zhang, C., Basta, T., Hernandez-Lagunas, L., Simpson, P., Stemple, D.L., Artinger, K.B., and Klymkowsky, M.W. (2004). Repression of nodal expression by maternal B1-type SOXs regulates germ layer formation in *Xenopus* and zebrafish. *Dev Biol* 273, pp. 23–37.

Zhang, C., Basta, T., Jensen, E.D., and Klymkowsky, M.W. (2003). The betacatenin/VegT-regulated early zygotic gene Xnr5 is a direct target of SOX3 regulation. *Development* 130, pp. 5609–24.

Zhang, J., Talbot, W.S., and Schier, A.F. (1998). Positional cloning identifies zebrafish one-eyed pinhead as a permissive EGF-related ligand required during gastrulation. *Cell* 92, pp. 241–51.

Zhang, Q., Yaniv, K., Oberman, F., Wolke, U., Git, A., Fromer, M., Taylor, W.L., Meyer, D., Standart, N., Raz, E., et al. (1999). Vg1 RBP intracellular distribution and evolutionarily conserved expression at multiple stages during development. *Mech Dev* 88, pp. 101–6.

Zhang, Y.Z., Ouyang, Y.C., Hou, Y., Schatten, H., Chen, D.Y., and Sun, Q.Y. (2008). Mitochondrial behavior during oogenesis in zebrafish: a confocal microscopy analysis. *Dev Growth Differ* 50, pp. 189–201.

Zhang, W., Poirier, L., Diaz, M.M., Bordignon, V., and Clarke, H.J. (2009). Maternally encoded stem-loop-binding protein is degraded in 2-cell mouse embryos by the coordinated activity of two separately regulated pathways. *Dev Biol* 328, pp. 140–47.

Zhao, X.D., Han, X., Chew, J.L., Liu, J., Chiu, K.P., Choo, A., Orlov, Y.L., Sung, W.K., Shahab, A., Kuznetsov, V.A., et al. (2007). Whole-genome mapping of histone H3 Lys4 and 27 trimethylations reveals distinct genomic compartments in human embryonic stem cells. *Cell Stem Cell* 1, pp. 286–98.

Zheng, P., and Dean, J. (2009). Role of Filia, a maternal effect gene, in maintaining euploidy during cleavage-stage mouse embryogenesis. *Proc Natl Acad Sci U S A* 106, pp. 7473–78.

Zhong, Z.S., Huo, L.J., Liang, C.G., Chen, D.Y., and Sun, Q.Y. (2005). Small GTPase RhoA is required for ooplasmic segregation and spindle rotation, but not for spindle organization and chromosome separation during mouse oocyte maturation, fertilization, and early cleavage. *Mol Reprod Dev* 71, pp. 256–61.

Zhou, Y., and King, M.L. (2004). Sending RNAs into the future: RNA localization and germ cell fate. *IUBMB Life* 56, pp. 19–27.

Zimmermann, J.W., and Schultz, R.M. (1994). Analysis of gene expression in the preimplantation mouse embryo: use of mRNA differential display. *Proc Natl Acad Sci U S A* 91, pp. 5456–60.

Zuccotti, M., Piccinelli, A., Giorgi Rossi, P., Garagna, S., and Redi, C.A. (1995). Chromatin organization during mouse oocyte growth. *Mol Reprod Dev* 41, pp. 479–85.